Chinese Research Perspectives on the Environment, Special Volume

Chinese Research Perspectives: Environment

International Series Advisors

Steven A. Leibo (*Russell Sage College*)
Li Yang (*Prop Roots Program*)

VOLUME 4

The titles published in this series are listed at *brill.com/cren*

Chinese Research Perspectives on the Environment, Special Volume

Annual Report on Actions to Address Climate Change (2012)

Edited by

Wang Weiguang, Zheng Guoguang, and Pan Jiahua

TUTA SUB AEGIDE PALLAS · 1683 ·

BRILL

LEIDEN | BOSTON

This book is the result of a co-publication agreement between Social Sciences Academic Press and Koninklijke Brill NV. These articles were selected and translated into English from the original《应对气候变化报告 (2012 气候融资与低碳发展)》(Yingdui qihou bianhua baogao 2012 qihou rongzi he ditan fazhan) with financial support from the Chinese Fund for the Humanities and Social Sciences.

This publication has been typeset in the multilingual 'Brill' typeface. With over 5,100 characters covering Latin, IPA, Greek, and Cyrillic, this typeface is especially suitable for use in the humanities.
For more information, please see brill.com/brill-typeface.

ISSN 2212-7496
ISBN 978 90 04 27463 1 (hardback)
ISBN 978 90 04 27464 8 (e-book)

This book is printed on acid-free paper.

FSC www.fsc.org MIX Paper from responsible sources FSC® C109576

Printed by Printforce, the Netherlands

Contents

Foreword

One of the great ironies of the struggle to confront the challenges of climate change is that while its origins are deeply tied to the evolution of the Western world's historical development, the key to confronting the challenge will most probably emerge from the developing world. Indeed, the reality of the situation becomes obvious when one considers the history of humanity's having unwittingly disrupted the global climate system with its prolific burning of fossil fuels.

It is, after all, well known that the more than two centuries of burning coal and other fossil fuels, including petroleum and natural gas, to generate energy was, until recently, largely a Western phenomenon. In fact, given the relatively long life of greenhouse gases, which can stay in the atmosphere for more than a hundred years this is more than historically interesting, it is literally true that the fossil fuel emissions of Western societies from generations long departed are currently playing a decisive role in determining both the fate of their own descendants and descendants yet to be born throughout the globe.

It is of course true as well that significant contributions to our understanding of global warming have been made Western scholars, from the first efforts by the mid-19th-century French scientist Joseph Fourier, who realized that the earth should have been colder than it was, to his Swedish counterpart working later in that century, Svante August Arrhenius, who studied the potential impact that the progressive increase of carbon dioxide (CO_2) in the atmosphere might have on global temperatures. Not forgetting more recent scholars and researchers from Roger Revelle and Charles David Keeling's work on rising CO_2 levels, as well as contemporary researchers like NASA's James Hansen and Pennsylvania State University's Michael Mann, who have all done yeoman's work to help develop a global understanding of global climate instability caused by generations of fossil fuel burning.

As is also well known, the United States led the way in the environmental movement of the 1970s by raising global consciousness of the increasingly grave effect humanity's growing economic accomplishments were having on the environment. Sadly, more recently, due to ideological differences and the generous spending of the domestic fossil fuel industry, the United States has fallen behind in confronting the reality of climate change, but even as that has occurred, the European Union has taken up the slack with its impressive commitment not only in to green energy development (i.e., solar energy in Germany) but also to a broader leadership role within international climate change negotiations.

Be that as it may, the last twenty years have seen the emergence of a globalized world economy; a phenomenon largely resulting from the collapse of the ideologies of socialist-style command economies and India's horrendously convoluted Regulation

Raj, which, as is well known, was conveniently coupled with the explosion of communication and transportation tools associated with the Internet and global delivery networks like FedEx, both of which have brought the world's two largest societies, India and especially China, into the planet's fossil fuel-driven, consumption-based industrial society.

Given that reality, it cannot be a surprise that though the world's CO_2 levels that hovered around 280 parts per million (ppm) before the industrial revolution, and it had only risen to 354 ppm by 1990, it is now over 400 ppm. This growth began to skyrocket especially in the last twenty years when India and China entered the world economy, each with a deep commitment to improve the economic lives of its citizens.

Unfortunately, the truth is that if these countries' understandable commitment to modernity had followed the Asian telecommunications model of simply skipping the cumbersome, older, Western landline (the telephone pole stage), leapfrogging into the world of individual mobile phones would probably not have been terribly problematic. But, in contrast to the choices made in telecommunications, on the level of energy, Asia's giants have chosen to follow more precisely the Western historical model of energy generation by burning their way through fossil fuels.

That decision, perhaps the most profound of modern world history, has had far-reaching implications. In China's case, the nation was able to single-handedly pass the previous champion of annual atmospheric emissions, the United States, in 2007—and that was only at an early stage, when neither India nor China had the majority of their citizens using energy at anywhere near American levels.

Thus, ironically, the emergence of Asia's giants, especially China, as full-fledged members of the globalized world economy absolutely guarantees that the long-term planetary path of CO_2 production, as well as solutions to that challenge, will not be based where it all began, in the West, but instead within Asia, most probably the result of decisions made today in China during this, the first decade, of the 21st century.

Given these realities, a better understanding of how developing nations, and especially China, view the challenge of anthropomorphic climate change cannot be considered anything but absolutely critical for humanity's future, testifying to the importance of Brill's English-language translation of the *Annual Report on Actions to Address Climate Change (2012): Climate Finance And Low Carbon Development*, which stands as an important contribution, along with its parallel publication, the *Annual Review of Low Carbon Development in China (2011–2012)*, as a document that seeks to rectify the extraordinary confusion found in the West on China's green energy future and its larger perspectives on this extraordinarily crucial topic. More specifically, the individual chapters of the *Annual Report on Actions to Address Climate Change (2012): Climate Finance and Low Carbon Development* focus on the following topics.

In international political negotiations on tough issues, it often takes decades of effort before one sees any progress. The 2012 Doha conference is already twelve months in the past, but it's still very interesting and useful to read the article by Wang, Pan and Lian, forming Chapter One of this volume, focusing on the Durban Platform created in 2011. The article gives a bigger picture of international negotiations on climate change.

The 2011 Durban conference was a turning point, both for international climate negotiations and for China's performance at those negotiations. Since the Durban conference, two mandates and three tracks of international climate negotiations progressed, simultaneously. The negotiation structure itself reflects the complexity of the problems. The challenges for China were greater than ever. People started to question: If even the hard-fought existing legal frameworks (the Kyoto Protocol and the Bali Roadmap) are not well implemented and finally abandoned, how can we trust the newly formed Durban platform? But trust and confidence in the multilateral mechanism are key to reaching a practical agreement and combating climate change. It's good to see that China's attitude has been evolving from its mainly defensive position at the Copenhagen conference to a much more active and mature strategy in international climate negotiations, and it will play a leadership role, together with other major players. However, the real challenge China faces, beyond mindset and negotiation skills, is balancing international expectations with domestic development needs, which is the core issue discussed throughout this volume.

The traditional clear-cut North-South coalition becomes obscured in international climate negotiations. Chao and Jia discuss newly emerging negotiation powers, and newly formed interest groups. Given the similar vulnerability of developing countries, South-South cooperation in coping with climate change has become even more crucial, especially on adaptation and setting up market mechanisms for emission mitigation without harming economic development.

This article not only reviews the history and development of South-South cooperation against climate change, but it also discusses cooperation actions that are already undertaken by China and other developing countries, such as technical and financial support for one another. Challenges to achieving more effective South-South cooperation exist, and more innovation is needed. The authors analyze the existing problems and also give concrete suggestions, such as setting up a special fund for South-South cooperation against climate change.

Chapter Three starts with a general theme, giving a full introduction to the background and global context of sustainable development and a complete list of China's national actions on sustainability. It follows with further discussion on transitions, opportunities, and uncertainties.

The discussion of China's specific outlook provides concrete suggestions on institutional reform, policies, and how to create a fair environment. It outlines steps toward a comprehensive transformation, from increasing the transparency of resource and energy demand to making better environmental risk assessments, as well as from focusing on the key driving forces and creating policy incentives to changing China's foreign economic cooperation strategy and enhancing corporate social responsibility for overseas development.

Chapter Four discusses aspects of the energy saving and low-carbon development targets during the 12th Five-Year Plan Period. China's 11th Five-Year Plan (FYP) (2006–2010) outlined, for the first time, a quantitative goal of reducing China's energy consumption per unit of gross domestic product (GDP) by 20% (i.e., in 2010, the energy consumption for generating each Yuan of GDP should be 20% below the level of 2005). The national goal was then disaggregated into provincial goals, differentiated according to their different social-economic situation. The actual reduction achieved was 19.1%, very close to the 20% target. Some may still remember how crazy some provinces were at the end of 2010, using blackouts as their final resort to meet their emission reduction goals. This was definitely against the intention of the national targets, but it also shows how serious the targets were and how challenging it was for some provinces to shift to a low-carbon economic structure within such a short period of time.

With fewer lower hanging fruits left, the 12th FYP (2011–2015) is even more challenging. Both central and local governments must be ready to make deeper changes in their patterns of economic development. The 12th FYP targets reflect China's promise to the international community and its domestic, strategic needs to promote economic restructuring and the transition of growth patterns. This article explains how the national targets are allocated to each province as well as the major challenges and the specific initiatives to achieve the targets. It also gives policy recommendations and points out the mismatches (for instance, the mismatch between the GDP growth rate target and the total energy consumption control target at the national level), which is an interesting phenomenon in China's policy systems that is due to not enough coordination or interest conflict between different sectors or government agencies.

As Zhuang and Wang describe in Chapter Five, since the 11th FYP period (2006–2010), when many provinces and cities started trying to get rid of their energy-intensive images, "low-carbon city" has become a buzzword, a sexy branding tool, among more and more Chinese cities. Starting with sporadic attempts in some cities, low-carbon city development policies have now been formalized as strategic plans at the national level. The concept of a "low-carbon city" has been interpreted as having very specific and substantial actions, but it's still at an early trial stage in China. This article introduces key national low-carbon city initiatives and analyzes different models through existing cases.

Chinese national policy is typically implemented by first reviewing experiences from pilot demonstrations and then promoting the policy nationwide, but the promotion process has never been smooth. As discussed in this article, the "Five Provinces and Eight Cities" program initiated by the National Development and Reform Commission (NDRC) is a serious high-level plan to make profound changes and spread the outcome around the country. The article also gives valuable suggestions for ensuring a successful national low-carbon city program, from research and urban planning to the monitoring and evaluation of emission reductions.

Regardless of the uncertain fate of the international Clean Development Mechanism (CDM), domestic market mechanisms are widely considered an indispensable way of reducing carbon emissions. As Chen, Lin and Zhou discuss in Chapter Six, China is taking ambitious steps in developing its own policies on carbon trading and will gradually establish its own domestic emission trading market. All the progressive initiatives discussed in this article are positive, valuable attempts at establishing China's carbon emission trading system. But it must be noted that many infrastructural facilities for carbon market operation are needed, among which the greenhouse gas (GHG) emission accounting system plays a key role. Given past experiences and the current situation in China, both addressing technical barriers as well as preventing corruption in the system are major challenges to establishing an effective GHG emission accounting system. Such an accounting system is crucial for ensuring that the whole market mechanism is contributing to real carbon reductions.

As the number one emitter of greenhouse gases, China is also one of the biggest victims of climate change and is facing the dual challenges of mitigation and adaptation. The article by Wang and Zheng summarizes the literature and practices regarding collaborative management of mitigation and adaptation actions and concludes from its research that about half of mitigation and adaptation actions could produce synergistic effects. The article also identifies some key fields and discusses several scenarios for optimization of synergistic effects. The article's analysis of the real case of Guangdong Province helps elaborate on the above theoretical discussions.

Accounting for 61.2% of global oil consumption, transport has become the sector with the largest and fastest-growing level of oil consumption, worldwide. This is even more the case in China, given its rapid growth in motor vehicle ownership. The article by Cai and Feng starts by accounting for CO_2 emissions released from China's transport sector then analyzes the relationship between CO_2 emissions in the transport sector and socio-economic factors.

In this article's analysis of the driving forces behind Chinese transport sector CO_2 emissions, the strong correlation between transport CO_2 emissions and regional GDP generally demonstrates that CO_2 emissions from the transport sector in China are mainly driven by the intensity of production activities rather than consumption activities. This also implies the great potential for consumption transportation emission growth (such as air travel generated by the tourism sector) along with improvements in living conditions. The article provides concrete policy recommendations based on comparative studies and statistics, which serve as solid reference materials for Chinese policy makers.

In the past three decades, China has been making great efforts to evolve from the "planned mandate economy" to a free market economy and is still struggling with the remaining shackles of the old system. Therefore, many basic facilities and the regulations for China's financial system are still in their early developmental stages. Too much reliance on public funding and difficulties behind removing barriers to private sector participation are common to many social and environmental issues, including the carbon emission reduction projects.

Research findings from the article by Pan, Chen, Yu and Wang clearly show that the costs of China's emissions reduction are growing year by year and will reach 2.47% of GDP by 2020. It is impossible to deny the trend in large-scale private investment and financing for low-carbon development as well as in many other traditionally state-dominated sectors. These are clear indicators of the urgent need for China to make substantial reforms in its financial system and economic structure.

Chapter Ten describes the innovation and practice of China's Green Climate Fund (GCF). This fund was established at the 2011 Durban Conference, but without enough detailed arrangements to make it operational. The China Green Carbon Foundation (CGCF) is an encouraging innovation and exploration into financing domestic carbon reduction projects through afforestation-related carbon offsets and carbon neutral programs. The Fourth Assessment Report by the Intergovernmental Panel on Climate Change (IPCC) states: "Forestry has multiple benefits, including both climate change mitigation and adaptation, and has been acknowledged as an important low-cost and economically viable measure in the next 30 to 50 years."

It is worth noting that CGCF is a professional non-profit organization founded and operated by China's State Forestry Administration. This ensures its technical ability and authority to implement national scale programs. Unlike the stereotypical way most Chinese government aid agencies operate, CGCF is very active in partnering with the private sector and very keen on promoting its projects among the general public. Although the accounting and verification of the effectiveness of its carbon offset projects still requires a high level of transparency and accuracy, CGCF has been exploring a way that can work well with the Chinese local situation and is leading a positive trend in China to create multiple innovative channels for financing carbon reduction projects.

Chapter Eleven discusses weather index-based insurance and its commercial application in China. China is a country that never lacks innovation and creativity, but the way to connect them with the market is always full of barriers, and there are not enough basic facilities and policies to support such an initiative. Given the tremendous loss caused by China's extreme weather conditions and the increasing frequency of such cases, weather index-based insurance has been widely discussed in Chinese academia and its business sectors. This article compares weather index-based insurance with the traditional property and casualty insurance and presents various problems in developing a weather index-based insurance product and a market for it under China's current situation, as well as ways for healthy operation and regulation. The article also shares a case study conducted by the China Meteorological Administration in Fujian Province. It gives a detailed analysis of the local weather and other natural conditions, the risks the tobacco farming industry is facing, and the design for an insurance product in response to these risks. The article ends with policy recommendations, including the integration of the national disaster response system and risk management measures with the development of a weather index-based insurance market.

In Chapter Twelve, Wang and Zhang discuss the development status and market prospects of renewable energy and new energy in China. In the past five years, China overtook the United States by becoming the largest emitter of GHG, but in 2009 and 2010, it also overtook the U.S. in its total investment in renewable energy. In 2010, China invested $45 billion in wind power (more than the entire U.S. clean-energy economy), which led to 17 gigawatts of new installations (more than three times those installed by the U.S.); however, the United States rose to the top again in 2011, and the world's total clean energy investments continue to increase to record heights. Countries seem to be more aware of the fact that even for sheer economic reasons, they cannot afford to lose in this "clean energy race."

China has committed to the international community to increase its share of non-fossil fuels (including nuclear) to 16% of the country's final energy mix by 2020. By 2012, it has achieved 9.1%, after developing at an unprecedented speed for almost a decade. Given the constrains on maintaining that speed and the strong resistance from coal and oil industries, achieving the 16% goal by 2020 remains a big challenge for China.

This article goes through the early development and current status of renewable energy and new energy in China since the passage of China's Renewable Energy Law in 2005. It compares different types of non-fossil fuels (including hydro, wind, solar, biomass, and nuclear) based on their technical maturity and efficiency in terms of GHG reduction. It also discusses the existing and needed incentive policies and regulation systems for renewable energy development and provides outlook and expectations for their future in China.

In Chapter Thirteen, Zhu Rong discusses wind and solar energy resources and their development potential in China. Spending $52 billion on renewable energy in 2011, China was responsible for almost a fifth of the world's annual total investment volume. The past decade is regarded as the "Great Leap Forward" in China's renewable energy development, especially for wind and solar power. Each year, about half of the world's new solar panels are manufactured in China, and China's total installed capacity for wind power generation has doubled every year for more than four years. This all presents huge challenges for the country's infrastructure, market system, and other facilities to keep up with this speed. Wind farms are having a difficult time connecting to the transmission grid, and the investments in wind and solar seem "overheated."

China's largest grid operator has committed to spending $88 billion on ultra-high-voltage transmission lines by 2020. By 2010, 31 out of 41 gigawatts of national wind installations were connected to the grid. The Chinese government has also committed to improve the connection situation and issue regulations for national wind power operations in the near future. All these problems and bottlenecks are to be expected during the process of development. It is quite encouraging to see China's determination to maintain this fast pace of development in its wind and solar power.

This article introduces the current utilization of wind and solar energy resources in China. Resource distribution and development costs are also discussed. It also takes a look at the facilities, incentives, and regulation policies, such as feed-in tariffs, cost allocation, development of smart grids, distributed access, and so on, which should be in place to ensure the healthy development of wind and solar energy in China.

As a large country with diversified eco-systems, China has experienced almost every type of meteorological disaster: droughts, floods, tropical cyclones, heat waves, chilling damages, frost and snowfall, sand-dust storms, etc. For nearly 20 years, these meteorological disasters have resulted in increasing direct economic losses and loss of life in China. It is estimated that meteorological and secondary disasters in China will continue to increase over the next 10 to 50 years. The article by Gao, Zhao and Xu in Chapter Fourteen provides detailed analysis for the economic and social impact of each type of meteorological disaster and looks at the specific impact in high, medium, and low greenhouse gas emission scenarios. It claims that social and economic development has contributed to a higher vulnerability of human society. Policy recommendations are integrated in all the above discussions.

Research papers on China and climate change issues are overwhelmingly focused on mitigation and clean energy development, whereas studies on adaptation, risk assessment, and management are not as frequent and very often not as thorough in their examinations. The article by Xu, Xu and Feng analyses the extreme climate events in China, using simulation tools from the five global climate models constructed by the Intergovernmental Panel on Climate Change (IPCC). The research results clearly indicate the accelerating trend of extreme climate events, due to the additive effect of the changes in their frequency, strength, spatial distribution, and duration. As a result, more and more people and facilities are exposed to higher risks from extreme climate events.

China is not a stranger to natural disasters, but the current risk management and emergency response systems are not of the same scale as the protection needed for its people, according to the studied scenarios. This article can serve as an important warning and useful reference for Chinese decision makers.

The article by Cao, Su, Zhai, Gemmer and Zhan covers an interesting social-economic aspect of the climate change issue. The world's most populated and economically developed areas are often along the coastlines. The rising sea level and extreme weather conditions combine to make the coastal cities more vulnerable than ever, and the potential loss is unprecedented. Unfortunately, city planning and development strategies in coastal areas still do not adequately account for these factors.

The article lists and compares the adaptation measures by Chinese and foreign coastal cities such as Shanghai, Guangzhou, Tianjin, Xiamen, Hong Kong, New York, London, Tokyo, Sydney, Venice, Singapore, and Cape Town. It points out that rapid urbanization aggravates deficiencies in infrastructure, and therefore, increases the challenges developing countries are facing. It ends with an analysis of climatic risk management in coastal cities in China and gives practical suggestions on actions to take.

Steven A. Leibo and Li Yang

CHAPTER 1

Doha Planning for the Future—Extension of the Kyoto Protocol and Outlook for the Durban Platform

Wang Mou, Pan Jiahua and Lian Huishan

Abstract

Two mandates and three international climate negotiation tracks have been undertaken simultaneously since the Durban Conference. This negotiation structure reflects the complex dynamic of negotiations in the climate change arena. Developed countries intend to conclude the two-track negotiation under the Bali Mandate and have been proactively promoting negotiations utilizing the single platform model of the Durban Platform. Developing countries seek further clarification regarding post-2012 Green House Gas (GHG) reduction obligations and responsibilities developed countries will face as part of the Long-Term Cooperative Action Working Group, as well as further information on the extension of the second commitment period of the Kyoto Protocol formed by the Kyoto Protocol Working Group. This three-track negotiation dynamic unveils the deeply divergent interests of the Parties. As different interests reassemble for climate negotiations, the traditional, clear cut North-South coalition becomes obscure. New negotiation powers are emerging and new interest groups have formed. These developments will no doubt change the international political hierarchy of climate change related decisions. Compared with the other two tracks, the Platform has received much more international attention due to its introduction of a new page of international

* About the author: Dr. Wang Mou, associate professor, is engaged in research on the international climate regime, environmental governance, green low-carbon development and other related issues, such as long-term tracking and participation in the international climate negotiations process. Funded projects: National Social Science Fund (12CGJ023); China Clean Development Mechanism Fund Grant Project (1112097). Pan Jiahua is general director at the Institute for Urban and Environmental Studies, the Chinese Academy of Social Sciences (CASS), research fellow and doctoral supervisor, specializing in world economics, climate change economics, urban development, energy and environmental policies, etc. Lian Huishan is working for the International Finance Corporation (IFC) of the World Bank Group. She is An Access to Finance Operations Analyst of the China Energy Efficiency Finance Program.

 | DOI 10.1163/9789004274648_002

climate negotiations As a result, ever party to the negotiations is attempting to establish a solid foundation for future talks under this track. Important issues such as agenda (deciding the chairperson of various working groups, negotiation a time frame for the Durban Platform etc.), principles of the new agreement, legal forms, frameworks and a roadmap of the Durban Platform will be of primary concern to interested parties during the Doha Conference. Under the Durban Platform, China will face greater international pressure to reduce its GHG emissions. Due to China's rapid economic development, the international community has increasingly higher expectations of China, asking it to assume more responsibility in addressing climate change. However, hurdles to further social and economic development alongside challenges to emissions reductions remain prominent in China. Therefore, China is expected to face great pressure to balance international expectations and domestic development needs at the Doha Conference. In order to achieve an international climate agreement that benefits the global environment, negotiations should be pursued under a just and cooperative framework that complies with the principles of the convention and acknowledges differing national realities.

Keywords

climate negotiations – Kyoto Protocol – Durban Platform – Doha Conference

The year 2012 marks an important landmark in the history of international climate negotiations. The first commitment period of the Kyoto Protocol will expire at the end of 2012; the Durban Mandate negotiations have already started; the negotiation mandate of the Bali Action Plan, which was originally scheduled to conclude in 2009, is also gradually coming to an end. International climate negotiations in 2012 are taking place under a complex landscape where two negotiation mandates are occurring simultaneously, and an open debate has been opened between nations over the outcome of the Durban Conference.

I Major Achievements of the Durban Conference

The Durban Conference continued to promote two-track negotiations using the Bali Action Plan and the Cancun Agreements, and negotiators reached a compromise during the conference with regard to the second commitment period of the Kyoto Protocol and the Durban Mandate. Parties hold different understandings of the time frame of the Durban Mandate (some countries argue that the Durban Mandate should focus the negotiations on a post-2020 international climate regime, while others insist that the Durban Mandate should cover emissions reduction efforts both before and after 2020). Once the new mandate comes into effect, it will gradually replace the Bali Mandate.

A *Extension of the Second Commitment Period of the Kyoto Protocol*

After long negotiations that ran far over time, parties reached a package of agreements that made the Durban Conference an historic turning point for international climate negotiations. Developing countries, including China, continued to demand a second commitment period for the Kyoto Protocol, and received support from the EU. The extension of the Kyoto Protocol into a second commitment period signifies developing countries have achieved a certain degree of success in the implementation of the "Bali Roadmap." The EU has agreed to implement the second commitment period of the Kyoto Protocol, and played a pivotal role in its promotion. Although the EU's emissions reduction targets are very conservative, it is nevertheless a symbolic victory for developing countries. Meanwhile, developing countries have not taken a rigid position that would require a 25%–40% emissions reduction of the EU by 2020,[1] which is the estimated emissions reduction target for developed countries according to the IPCC Assessment Report. It reflects some flexibility on the part of developing countries, which made a vital contribution to the success of the negotiations. The United States did not join the first commitment period of the Kyoto Protocol, and Canada, Russia, and Japan have joined the US in refusing to endorse the second commitment period. As such, the US has failed to lead developed countries in addressing climate change during the negotiations. Nevertheless, the EU's position on the Kyoto Protocol is worthy of appreciation and recognition.

B *Onset of the Durban Negotiation Mandate*

Developing countries made an important compromise regarding the manner of initiating the Durban Mandate. They agreed to start with the Durban Platform negotiations that cover all nations' emissions reduction actions and commitments and to reach an emissions reduction agreement by 2015. The Durban Mandate is a new negotiation mandate that covers emissions reduction commitments of all major emitting countries.[2] The EU had aspired to

---

1 IPCC, *Climate Change 2007: Mitigation*, (Cambridge: Cambridge University Press, 2007).

2 Draft decision -/CP.17. Establishment of an Ad Hoc Working Group on the Durban Platform for Enhanced Action, http://www.unfccc.int/.

develop the new mandate before the conference, and it was very pleased with the outcome in Durban.[3] Since 2009, a number of developed countries have proposed to put aside the differences between developed and developing countries and abandon the Kyoto Protocol so that all Parties can negotiate under the same framework. Under the Durban Platform their goals were also partially achieved.

C *Progress was Made in Green Climate Fund Development, Technology Transfer, Adaptation and Other Reduction Mechanisms*

Although negotiation Parties could not achieve a consensus on the funding sources for a Green Climate Fund, they consented to terms such as who will make up the board of the Fund, what their responsibilities will be etc. Progress on technology and adaptation issues was a target of developing countries, and the negotiation Parties clarified the framework regarding these issues during the Durban Conference. Future negotiations will further refine the details surrounding these issues.

The Durban Conference has propelled international coordination on climate change. The cooperation between the EU and developing countries was a key to its success. Because the United States did not receive the necessary authorization from the US Congress, it was difficult for the US to make positive contributions to the conference's success. Other umbrella countries, such as Russia, Canada, and Japan, were not enthused by the progress that was made, but they did not block the negotiation process, which is a contribution in it of itself, considering their relative weight. The Durban Agreement is a compromise between a number of negotiating powers, and just that can be recognized as a success for climate negotiations.

II Major Issues in the Durban Platform Negotiations

The Durban Platform, in its role as a new negotiation mandate, must reflect the negotiation interests of all associated Parties, and there is no doubt that it will attract a great deal of attention. Procedures, principles, legal forms and other issues of the Durban Platform form the primary discussion at this stage.

---

3 When Friis Arne Petersen, Danish Ambassador, called the author at Institute for Urban and Environmental Studies of the Chinese Academy of Social Sciences on December 16, 2011, he said the establishment of Durban platform is the "victory of the climate."

As the negotiations develop, they will gradually be able to expand to other more specific issues.

A *Time Frame of the Durban Platform Negotiations*

Due to vague language in the Durban Platform mandate resolution, negotiation Parties have been unable to reach a consensus on the time frame of the Durban Platform negotiations. Certain parties hold that the Durban Platform should focus the discussion on the post-2020 international climate regime, and that the work to intensify emissions reduction efforts before 2020 should be handed over to the Long-Term Cooperative Action Working Group and the Kyoto Protocol Working Group which operate under the Bali Mandate. The EU and AOSIS contend that the Durban Platform negotiations should cover the contents of both time periods, namely, how to enhance global emissions reduction efforts before 2020 and how to develop the international climate regime after 2020. This position may render the Bali Mandate a de facto extension of the Durban Mandate because of overlap between core issues of the two working groups, such as the post-2012 emissions reduction targets and emissions reduction action targets. If these issues are taken up by the Durban Platform, it will hinder momentum for negotiations as part of the Bali Mandate.

B *Principles of the New Agreement*

Principles are indispensable for any agreement. They are essential for designing the negotiation framework and guiding the negotiation process. The Durban Platform is a negotiation mandate under the United Nations Framework on Climate Change Convention, thus, its principles should be in line with the principles of the Convention. The primary principles the Parties disagree on are connected to the concept of "common but differentiated responsibilities (CBDR)." Developing countries generally believe that the Kyoto Protocol is an appropriate interpretation of the CBDR principle, meaning they are willing to commit to a total emissions cap and offer financial and technical assistance to developing countries so as to help them improve their ability to adapt to climate change. Under this interpretation, developing countries should undertake poverty alleviation and economic development as their priorities, and take action to reduce greenhouse gas emissions based on their respective capabilities. Despite the general consensus, a number of developed countries have proposed that as the global economy develops, the international community should develop a dynamic understanding of the CBDR principle and developing countries should take on greater responsibilities for emissions reductions. In addition, a few developed countries refuse to recognize the CBDR principle.

Great differences in understanding the CBDR principle persist among Parties. The definition of the CBDR principle and its use in guiding the negotiation process of the Durban Platform are key issues that must be addressed in during the Durban Platform negotiations.

C *Legal Structures*

The legal structures of future climate agreements have comprised one of the primary focuses of the international community in recent years. Before the Durban Conference, the legal structure issue centered on the potential agreement under the LCA working group. But since the negotiations of the Durban Platform Mandate have kicked off, the international attention has drawn to the legal structure that will arise from the Durban Platform. The EU and AOSIS support an agreement that legally binds all parties. India, China, and other developing countries refuse to determine the legal structure of the agreement at this stage, contending that there are still too many uncertainties regarding the socioeconomic development level after 2020. These differences regarding the legal structure of the Durban Platform agreement are likely to persist, and in doing so evolve to be a primary issue during the Durban Platform negotiations.

D *The Durban Platform's Framework and Agenda*

The framework of the Durban Platform is an essential issue for future negotiations. The Durban Platform Parties must decide whether the framework shall be built on existing negotiating texts, such as Kyoto Protocol or LCA, or whether it shall be restarted from scratch. Difficult negotiations are expected regarding the agenda, as it needs to balance all Parties' interests. Furthermore, questions remain as to whether a new agreement should attempt to address issues that were not resolved under the Bali Mandate. One can presume that due to the limited capability of developing countries to predict their future socioeconomic development needs, they will adopt a conservative position on the post-2020 international climate regime during negotiations. They will be particularly cautious in addressing the framework and agenda issues of the Durban Platform.

E *Roadmap and Time Frame*

The Durban Platform Mandate has stated that Parties should complete negotiations before 2015. This statement has set an undoubtedly short time frame for Parties to work with. A series of disagreements between Parties remain regarding emissions reduction targets, global emissions peak, finance and technology assistance, compliance methods, and other important issues. The domestic climate legislation in the US is lagging, and the political momentum of the IPCC's

Fourth Assessment Report has almost been exhausted. If no breakthroughs are made on these issues, it is difficult to imagine that the negotiations could be concluded by 2015.

III A New Pattern of International Geopolitical Climate Politics

A *Three Track Negotiation with a Focus on the Durban Platform*

Two mandates and three tracks of international climate negotiations are being simultaneously undertaken. Negotiations are being conducted under LCA, KP and the Durban Platform at the same time. This structure reflects the complex dynamic of the climate change negotiation arena. Under the joint promotion of the EU and some other developed countries, the Bali Mandate will gradually be replaced by Durban Mandate, and the focus of international climate negotiations will shift to the Durban Platform. The LCA working group Parties under the Bali Mandate will also shift their attention onto how to build a new framework for a post-2020 international climate agreement.

B *The North-South Coalition Obscured*

Since the Durban Conference, the traditionally clear cut North and South positions on key issues have become obscure. During the Conference, the EU, AOSIS, and Least Developed Countries Group published a joint statement on initiating the Durban Mandate. Deep divergences of positions on the Durban Mandate as well as different interpretations of the Mandate exist among developing countries as well as developed countries. During the negotiations held in Bonn in May 2012, Group 77+ China did not reach a consensus on the time frame of the second commitment period of the Kyoto Protocol,[4] which led to a negotiations stalemate. The EU aligns with AOSIS and appeals for a legally binding agreement with regard to the future climate agreement's legal structure. The US is taking a position similar to that of some developing countries, who do not want to determine the legal structure of future agreement at this stage. As different interests reassemble in the climate negotiations, the line between the North and the South becomes ever vaguer, resulting in further complications to the negotiation dynamic.

4 The implementation period of the second commitment period of the Kyoto Protocol is five years, and it will end in 2017 or in 2020 if the implementation period is eight years.

C *Reassembly of Major Negotiation Powers and Emergence of New Patterns in Climate Negotiations*

As Parties' negotiation positions diverged further, the traditional group coalition pattern began to split. New negotiation powers emerged and like-minded interest groups were formed. Despite these changes, the dominant role of the United States cannot be shaken, as it remains the hegemonic superpower politically, militarily, and economically. It is still among the primary negotiators in the climate negotiations, regardless how new pattern evolve, and whichever negotiation direction the US chooses, most Parties under the Umbrella Group will follow. The EU is facing a twofold challenge: failing manufacturing industries and a financial crisis. The total GDP of the EU's 27 members accounted for 27% of global GDP in 1995, but the proportion dropped to 23.9% in 2009, a decrease of 3.1% in 14 years. CO_2 emissions dropped from 17.7% in 1995 to 12.3% in 2009, a fall of 5.4%.[5] The EU's economic slowdown not only affects its leadership in international affairs, it also weakens the EU's influence on the UNFCCC negotiations after the 2008 Financial Crisis. As a result, the EU actively seeks to ally in negotiations in order to maintain a leadership role. The alliance between the EU, AOSIS, and the Least Developed Countries will persevere through the Durban Platform negotiations. It may be the largest like-minded coalition present at the Platform,[6] and become a major bargaining power in future negotiations. There is no doubt that the EU will play the leading role in this new coalition. In the negotiations after the Durban Conference, especially during the dialogue meeting in Bonn which took place in May, a group consisting of China, India, and other emerging economies have also gradually started to form a coalition mutual interests. These countries are experiencing fast economic development and fast emissions growth, and they find themselves in the critical juncture of economic development, poverty alleviation, quality of life improvement, and participation in the international trading system. They come together in order to avoid bearing unreasonable emissions reduction obligations and abatement costs, and to safeguard emissions space for equal development. This coalition will also be a major bargaining power in future climate negotiations.

D *Governmental Transitions Raises Uncertainties in Negotiations*

In 2012, more than 50 countries, including the United States, China, France, Russia, Mexico, South Korea, and Spain, will change their governmental

5 IEA, 2010: CO_2 Emissions from Fuel Combustion, 2010 Edition, P77.

6 A total of 120 countries, that is, 27 EU states, 43 small island states, 50 least developed countries, form a majority group among the 194 party states.

leadership. In the United States in particular, the Republican Party and the Democratic Party have different understandings on the climate change issue, and a change of the ruling party is likely to result in a change of the US' national position on climate change negotiations. The politically sensitive timing of elections and change of leadership in many countries may cause new environmental and climate initiatives to delay. Most countries with a general election will tend toward relative conservative stances and will attempt to maintain the existing environmental and climate policies. General elections can easily bear a serious influence on a country's climate change policy, which may add uncertainties to climate negotiations.

IV China's Dilemma in Durban Platform Negotiations

Curbing greenhouse gas emissions and reducing energy consumption are efforts important to China not only to address the climate change issue, but also to safeguard its energy security. China is willing to take voluntary emissions reduction actions despite an absence of international binding obligations under UNFCCC. China's energy intensity has decreased 19.1% under the implementation of its 11th Five-Year Plan,[7] and it experienced a decreases of about 1.5 billion tons of GHGs. China has been a proactive and constructive party in international climate negotiations from their inception, but due to its limited ability to predict future socioeconomic development and a series of challenges that its society and economy face, China cannot pursue unrealistic emissions reduction targets in climate negotiations. The following detail some of the hurdles to ambitious emissions reduction targets.

A *A Need for a Higher Level of Socioeconomic Development*

China remains at a relatively low level of economic development. National per capita GDP in 2010 was $4,430 (USD figure is calculated based on the exchange rate that year), 1/3 of the world average.[8] The regional gaps in economic development in China are also significant as is the income gap between urban and rural residents. Per capita disposable income of urban residents was $2,822 in 2010, but only $874 for rural residents, only 31% of urban residents' income. China also continues to face the serious challenge of poverty alleviation.

7 Xie Zhenhua, head of the Chinese delegation and deputy director of the National Development and Reform Commission, attended the high-level meeting of the Durban Climate Conference and delivered a speech.

8 2011 China Statistical Yearbook.

At the end of 2010, the poverty-stricken population in China's rural sectors numbered 26.9 million. Their average per capita annual net income was less than 1,196 yuan ($178).

B *Rapid Urbanization Has Led to Accelerated Emissions Growth*

China is currently undergoing rapid urbanization; the urbanization rate has risen from 31.9% in 2000 to 51.27% in 2011, an increase of 19.37% in 12 years. In accordance with the experience of the developed countries, the urbanization rate of a mature industrial economy rests at least at 70%. Based on the current annual urbanization growth rate of about 1%, China would not complete the urbanization process until 2030. Per capita energy consumption of urban residents is estimated to be 1.8 times that of rural residents.[9] The accelerated urbanization process will inevitably lead to increased energy consumption. Urbanization and the corresponding enlargement of the income gap will inevitably lead to the continuous increase of total energy consumption.

C *Industrialization and Emissions Transfer*

China's average annual economic growth rate has hovered around 10% for the last 30 years, and it has likewise been transitioning from labor-intensive to capital-intensive growth as a result of industrialization. China's crude steel production reached 683 million tons in 2011,[10] and cement production reached 2.06 billion tons,[11] accounting for about 50% of global production. China has also become the world's largest emitter of greenhouse gases in conjunction with its economic growth, according to some reports. This change is particularly acute in China's export sector, which has contributed significantly to its rapid economic growth. In 2011, exports accounted for 26% of China's total GDP.[12] Most of the intermediate products and consumer goods are shipped to markets in developed countries. According to current statistical rules, since the production of these export commodities takes place in China, their carbon emissions are accounted for entirely as China's responsibility. Some studies show that the energy consumed for the production of China's export commodities in 2006 accounted for 25.5% of China's total energy consumption that

---

9 2010 China Energy Statistical Yearbook.

10 China's crude steel production in 2011 was 683 million tons http://www.askci.com/news/201201/10/85451_38.shtml.

11 The annual national cement output in 2011 was 2.06 billion tons http://www.ccement.com/news/Content/49383.html.

12 National economic and social development statistical bulletin of People's Republic of China in 2011. http://www.stats.gov.cn/tjgb/ndtjgb/qgndtjgb/t20120222_402786440.htm.

year. These studies also predict that China's status as "the workshop of world" will remain unchanged in the coming years. Under those circumstances China will face enormous challenges in reducing GHG emissions in the long run.

D *Resource Endowments and Energy Structures are Difficult to Adjust*

China's remains dependent on coal at a level much higher than that of many developed countries. Coal accounted for 70% of China's total energy consumption in 2011, 30% higher than the world average. Oil and natural gas accounted for 18% and 5% of total energy consumption respectively, while nuclear energy and renewable energy including hydropower only account for about 7.4%, far below the 39.1% in France and the 12.1% in the world on average.[13] Coal is still a major energy source in China, and the coal dominated energy consumption pattern is unlikely to shift in the near future. China's limited resource endowment has largely limited its ability to reduce the carbon intensity of its energy sector. Due to the lack of advanced technologies and policies, including a lack of Intellectual Property Rights (IPR) regulations on key energy technologies, the investment and operation costs of new energy R&D are high.

E *Technology Lock-in Effect Lowers Efficiency*

The underdeveloped energy pattern and utilization technology is one of the main reasons China faces low energy efficiency and high carbon intensity. Compared with developed countries, China has many improvements to make to its technologies in energy development, energy supply and transmission, configuration, industrial production, and other technical aspects of energy end-usage. China's traditional industries continue to use outdated technologies as standard practice. Due to the lack of advanced technologies as well as the common application of outdated processes techniques, energy efficiency in China is currently about 10% lower than that in developed countries, and the energy consumption required per unit of energy-intensity products is about 40% higher than the standard set by international best practices.[14] As China undertakes large-scale infrastructure constructions in energy, transportation, and building construction, the usage of low-efficiency technology will result in as technology lock-in effect, locking China into a development pattern characterized by high energy consumption and low efficiency. This technology

---

13 BP Statistical Review of World Energy June 2010, http://www.bp.com/statisticalreview.

14 Chen Shihai: China's overall energy efficiency is 33%, 10% lower than that of developed countries. http://www.tianshannet.com.cn/energy/content/2009-02/27/content_3866541.htm.

lock-in effect poses a serious challenge to China and its efforts to address climate change and reduce GHG emissions.

F *China's Dilemma of Attempting to Meet High Expectations of the International Community and also Avoid Domestic Socioeconomic Development Constraints in the Durban Platform Negotiations*

China has been put under the international media spotlight in recent years due to its rapid economic development, increased total GHG emissions, and large foreign exchange reserves. The international community is requiring China to shoulder a greater share of GHG emissions reduction responsibilities as a result of pressure from certain countries. These requirements do not take the compelling needs of China's economic development, poverty alleviation, and improvement of quality of life into account, and they further ignore the challenges and difficulties in GHG reduction discussed above. To meet the unrealistic GHG reduction demands may become an unbearable weight that hinders China's social and economic development. China faces a dilemma at this stage of the Durban Platform negotiations, confronting both the urgent need for domestic socioeconomic development, and the strong pressure from the international community to reduce emissions. China is expected to receive a substantial amount of external pressure in the Durban Platform negotiations.

V Ideas for Promoting the Durban Platform Negotiations

The construction of the post-2020 international climate regime is of great significance for global climate governance. It is also an indispensable variable in global climate security. The Durban Platform negotiations must be built on the basis of mutual respect as well as just and equitable sharing of responsibilities and obligations, modeling previous international agreement negotiations. It should strive for mutual understanding and compromise, so that the negotiation process can progress efficiently.

A *Single-track Negotiations do not Implicate Identical Responsibilities and Obligations*

Climate change is the result of historical cumulative GHG emissions, and the contributions to GHG emissions throughout history are clearly different among countries. There is general consensus in international climate governance that developed countries must shoulder the major responsibility for climate change. Whether the international climate negotiations are carried out using a double-track or single-track approach is not the core issue, but rather different strategies to achieve a similar result. The different means should not be used to blur the distinct responsibilities and obligations of developed and developing countries. In single-track negotiations, although all parties are negotiating on a single platform, developed and developing countries should not be expected to shoulder identical international responsibilities and obligations.

B *Climate Justice in the Durban Platform Negotiations*

International climate governance deals primarily with sharing emissions reduction responsibilities and obligations among various Parties. Ongoing negotiations depend on a fair and equitable sharing mechanism. The Kyoto Protocol stipulates a total emissions cap on developed countries, and requires developing countries to carry out emissions reduction actions according to their respective abilities. The Protocol reflects the development needs of developing countries, and is in line with the consensus that developed countries, which contributed to the majority of historical GHG emissions, should adopt a leading role in reducing emissions. In so doing, the agreement accurately represents the spirit of international fairness and justice. However, some Parties have deviated from the common understanding of climate justice established in the Kyoto Protocol and have attempted to redefine the "common but differentiated responsibilities" principle in recent years. They advocate for developing and developed countries to make equivalent emissions reduction commitments and take equivalent mitigation actions. But is it fair for a state with an annual per capita income of $3,000 and per capita CO_2 emissions of three tons to make the same emissions reduction commitment as a state with an annual per capita income of $30,000 and per capita CO_2 emissions of more than 20 tons? If the Durban Platform negotiations cannot address the justice issue in the future climate regime, it will face a boycott from a widespread swath of the Parties, causing impediments of the international climate negotiations process.

C *It is More Effective for Developing Countries to Take Emissions Reduction Actions than to Commit to Emissions Reduction Targets*

The Durban Platform is a negotiation mechanism for the post-2020 international climate regime. The socioeconomic development levels, energy demand, and GHG emissions volume of developing countries that are undergoing unconventional development paths can only be predicted by positing hypotheses for the different scenarios. Due to the lack of predictability, even if the implementation timeline of the future climate agreement is extended to 2025, it will be difficult for any government to commit to GHG emissions reduction targets which it can only base on uncertain and shifting hypotheses.

With that in mind, it will be more constructive and effective to use the Durban Platform to support the emissions reduction actions of developing countries, rather than compelling them to accept overambitious and unrealistic emissions reduction targets.

D *Establishing Effective Financial and Technical Support Mechanisms*

To curb greenhouse gas emissions and to achieve low-carbon development requires not only the political will of the Parties, but also financial and technical support from the international community in the form of cooperation and assistance. Advanced technologies can improve energy efficiency and reduce energy use and greenhouse gas emissions, but technology acquisition, upgrades, and dissemination require huge investments. Take renewable energy as an example; the cost of generating wind and solar energy is far greater than the cost of using conventional energy sources such as coal, oil, etc. Achieving effective international cooperation on GHG reduction is dependent upon sharing climate-friendly technologies and a system that provides financial support and an economic safeguard.

VI Planning for the Future—The Historical Missions of the Doha Conference

The Doha Conference will function as a watershed for international climate negotiations. The first commitment period of the Kyoto Protocol is about to come to an end. Although certain elements of the second commitment period of the Kyoto Protocol have been largely agreed upon, Parties need to make clear and feasible goals and agreements for the second commitment period of the Kyoto Protocol at the Doha Conference, which is the last UNFCCC conference within the first commitment period. At the Durban Conference, the EU committed to the second commitment period of the Kyoto Protocol. The EU's 2009 climate action plan announced that EU will reduce its GHG emissions by 20% of 1990 levels by 2020. Although most developing countries have complained that the EU's emission reduction targets are not ambitious enough, they nevertheless recognize the EU's effort to continue the implementation of the Kyoto Protocol. Developing countries expect the EU to act as role model and encourage more developed countries to join the second commitment period of the Kyoto Protocol. They also seek a clear decision on the legal structure of the Kyoto Protocol's second commitment period in Doha. The negotiations of the Kyoto Protocol Working Group this year have reached a stalemate. AOSIS insists that the implementation time of the second commitment period should be five years and ended in 2017, while the EU proposes an eight year time frame, which extends the end date to 2020. This disagreement, along with other variables such as the legal structure, interim plans, and Parties' political commitments for the second commitment period of the Kyoto Protocol, cast further uncertainties over the international community's expectation for the Doha Conference. It appears that the Kyoto Protocol's future depends on AOSIS's position, but in reality the political wrestling taking place behind the scene is beyond the control of small island states. The key ultimately rests with the EU and other developed countries' resolution to take the lead and to substantially reduce their greenhouse gases.

As the Durban Platform launches in 2012, a greater proportion of Parties are supporting a conclusion of the Bali Mandate at the Doha Conference, namely the two-track negotiations of LCA and KP, and a focus shift onto negotiations under the Durban Mandate. The progress of the LCA negotiations reveal that disagreements exist on many grounds and consensus on most key issues has yet to be reached. There is currently insufficient international political will to conclude an agreement under the LCA track. An end to LCA negotiation at the Doha Conference under these circumstances will result in either a weak agreement with little binding force or a compromise that yields few substantial outcomes. Both scenarios are possible, but the latter is more likely. In this case, the substantial and controversial issues currently under LCA will likely be left for future negotiations, which may provide the opportunity to conclude the Bali Mandate at the Doha Conference. Once the negotiations under the Bali Mandate have come to an end, the Durban Platform negotiations will be fully underway, and international climate negotiations will shift to the next stage.

The historical mission of the Doha Conference is extend the Kyoto Protocol's second commitment period by the end of 2012, to bridge international negotiations from the Bali Mandate to the Durban Mandate, and to build a blueprint for the new mandate and for future negotiations. If it succeeds in accomplishing these objectives, the Doha Conference is destined to be another milestone in humankind's effort to address climate change and protect the future of younger generations.

(This article was originally published in Chinese in 2012.)

CHAPTER 2

South-South Cooperation against Climate Change and Future Prospects

Chao Qingchen and Jia Pengqun

Abstract

South-South cooperation plays an indispensable role in international multilateral cooperation. As globalization has spread over recent years, climate change has become a focal point of South-South cooperation and a force in negotiations and dialogues with developed countries that cannot be ignored. This article systematically analyses the role and status quo of South-South cooperation against climate change, providing a perspective on past experiences and existing shortages, as well a prediction of possible outcomes in the future.

Keywords

climate change – South-South cooperation – prospect

I The Role of South-South Cooperation against Climate Change

A *History of South-South Cooperation*

International cooperation can be broadly split into three categories: North-South cooperation, South-South cooperation, and triangular cooperation. North-South dialogue and cooperation refers to the negotiations, dialogues and multilateral consultation and cooperation on economic issues between developing countries and developed countries. Because most of the developing countries are situated in the southern hemisphere or the southern part of

* Chao Qingchen is the deputy director of the Beijing Climate Center, China Meteorological Administration (CMA). Her research area covers climate change data analysis, policy research and management.

Jia Pengqun is an associate researcher at the CMA Training Center. His research area is meteorological soft science.

 | DOI 10.1163/9789004274648_003

the northern hemisphere, the economic and technological cooperation among developing countries is called "South-South cooperation."

South-South cooperation began as the larger developing countries faced similar challenges and experiences upon obtaining independence. At the Bandung Conference convened in 1955, these countries agreed upon the principle of "consultation" for South-South cooperation, urged countries to develop organization for raw material production and exportation, and proposed capital and technological cooperation, marking the formal initiation of South-South cooperation. The Non-Aligned Movement and the Group of 77, established at the beginning of 1960s, are the two largest international organizations representing South-South cooperation. They have established a series of core documents that stipulate the general content, areas of interest, means of interaction, and basic guiding principles surrounding South-South cooperation. From the 1970s to the end of the 1980s, developing countries made significant progress through united, cooperative and self-reinforcing efforts. The West African Economic Community, Latin American Economic System, Southern African Development Community, Gulf Cooperation Council, South Asian Association for Regional Cooperation, and other developing countries all strove toward economic cooperation. Some regional economic organizations aimed at strengthening independence were set modelling one another. In 1978, the Conference of Technical Cooperation among Developing Countries was held in Buenos Aires, Argentina where the Buenos Aires Plan of Action was adopted. As a result, a strategic framework for South-South cooperation emerged which allowed these cooperative efforts to grow in impact and importance in dealing with world affairs. The first South-South Cooperation Conference was held in New Delhi, India in 1982 and the conference was reconvened in Beijing and Kuala Lumpur in 1983 and 1989, respectively. The three conferences are an important milestone for South-South cooperation. The United Nations appointed December 19th as South-South Cooperation Day in 2003 to increase awareness of cooperation efforts. The international community commemorated South-South Cooperation Day for the first time in 2004, with the theme of accomplishing millennium development goals. The principle purpose of South-South cooperation is to carry out joint self-improvement projects and promote mutual development while struggling under unequal North-South economic relations. This cooperation has become an indispensable part of international multilateral cooperation, an important format for developing countries to promote independence and progress, and an effective way for developing countries to participate in the world economic system.

South-South cooperation is mostly found in four arenas. The first is regional economic cooperation. Organization members reduce or eliminate tariffs for each other, implement free commodity circulation, create a standard tariff rate for non-member countries, and promote a common market. A second cooperation form is trade cooperation, whereby developing countries develop global trade preference systems and trade organizations. The third arena is monetary and financial cooperation. The Andean Development Corporation and Central Bank of West African States, for example, provide loans and aid to other member countries. Arab countries and some Arab monetary and financial institutions provide long-term, low-interest or preferential loans for developing countries without any additional political or economic conditions. The fourth platform is technological cooperation, such as technology transfer, patent sale, technological consultation and training, and technological information exchange. Countries also provide technical services and labor for joint ventures that lead to mutually beneficial technological advancement.

Global South-South cooperation tends to diversify development and strengthens the role of large developing countries. The increased cooperation begets further creative cooperation platforms and regional partnerships of various sorts. Developing countries have been able to improve the scope, depth, and diversity of cooperative projects even in the face of the great challenges of globalization.

B *Development of South-South Cooperation against Climate Change*

According to Article 4 of the *United Nations Framework Convention on Climate Change* adopted in 1992, all parties shall promote and cooperate on scientific, technological, technical, socioeconomic, and other research, systematic observations, and development of data archives related to the climate systems. Parties shall also promote and cooperate on the full, open, and prompt exchange of relevant scientific, technological, technical, socioeconomic, and legal information related to the climate system and climate change, and to the economic and social consequences of various response strategies. They shall further promote and cooperate on education, training, and public awareness related to climate change. The Kyoto Protocol, adopted in 1997, also emphasized cooperation in corresponding articles. However, it placed a greater emphasis on the responsibilities and obligations developing country members face in providing monetary support and transferring technologies to developing country members in order to aid in addressing climate change, increasing the efficacy of actions undertaken by developing countries. The focus remained on South-North cooperation.

As progress is made in climate negotiations, the international cooperation mechanism for addressing climate change continues to evolve. The Cancun

Agreement, adopted an the end of 2010, mentioned South-South cooperation in clauses that did not discuss the responsibilities and obligations of developed countries. For example, in discussion of "technology development and transfer," the Agreement stipulates stimulation and establishment of South-North, South-South and triangle cooperation opportunities and partnership in the arrangement and dissemination of existing technologies and in the establishment of national, regional, industrial, and international technical centers or networks. It further states that the Agreement parties shall enhance information and knowledge generation, sharing, and management through South-North, South-South and triangle cooperation in discussion of capacity building and aide to developing countries.

Roughly 30 years of South-South cooperation can be approximately divided into three development stages (10 years for each stage). In the first 10 year period, a common view was established through national focal points, which laid a sound foundation for taking action. In the second 10 year period, South-South cooperation shifted gears due to globalization. A huge number of projects and programs emerged under the banner of South-South cooperation. In the last 10 years, more substantial progress was made thanks to the growth of South-South trade, investment, and tourism. Today, South-South cooperation is an extremely important tool and structure for enhancing economic independence of developing countries and realizing sustainable development. It is also one of the most effective routes allowing fair access to the global economic system for developing countries. As climate change continues to be addressed in various platforms, the mechanisms governing South-South cooperation, and the content of agreements continues to evolve, becoming deeper and more complex. It is emerging as a force that cannot be ignored in negotiations and dialogues between developing and developed countries, allowing developing countries greater leverage to demand fairness and balance between demands to reduce carbon emissions and a protection of the right to development.

II Current Status of South-South Cooperation against Climate Change

The following section discusses current forms of South-South cooperation used to combat climate change, presenting efforts such as aid from emerging developing countries given to the most underdeveloped countries and small island countries and cooperation among emerging developing countries which aims to use each country's advantages to help address others' needs.

A Current Status of South-South Cooperation among Major Developing Countries

Many developing countries are deeply concerned about how to best adapt to climate change. The Korea Meteorological Administration (KMA) is one organization that puts forth great efforts to help African countries adapt to climate change. It contributed USD 1 million from 2009 to 2010 to support capacity building at the IGAD Climate Prediction and Applications Center (ICPAC) in the Greater Horn of Africa. It helped ICPAC perform the duties of a regional climate center, including making climatological observations, monitoring, offering predictions, disaster management, personnel training, etc. ICPAC has taken the lead on climate predictions and application services related to climate risk management, environmental management, and sustainable development of the seven member states in East Africa (Republic of Djibouti, Eritrea, Ethiopia, Kenya, Somalia, Sudan, and Uganda). At the Cancun Climate Change Conference, the environmental minister of India specifically emphasized India's South-South cooperation efforts in his speech made at the ministerial conference. India signed an agreement with Bangladesh to set up the Sundarbans Ecosystem Forum. The Sundarbans is the largest river delta in the world and is particularly sensitive to climate change. India has also subsidized the construction of the South Asia Forest Center in Bhutan as well as a coastal areas management center in the Maldives. Furthermore, it launched a capacity building and technical support plan for experts and scholars from small island countries and launched an ecosystem restoration plan with Nepal and China that seeks to help neighboring countries and the most underdeveloped countries to actively cope with climate change. The countries of China, India, Brazil, and South Africa (hereinafter referred to as BASIC) work with different arrays of economic growth and natural resources. However, since all BASIC countries are developing countries, they share a set of common interests at international negotiations on climate change and have both a need and a desire to cooperate in the fight against climate change. BASIC's minister-level climate change coordination mechanism has become an important platform for aligning differing positions in climate change negotiations. According to a repeatedly published joint statement, the mechanism is also a platform for cooperative action aimed at slowing down climate change effects and adapting to them, which includes information exchange and cooperation on climate science and climate related technologies. BASIC has already conducted extensive and fruitful cooperation efforts in the field of climate change. For example, in accordance with the framework of bilateral intergovernmental science & technology cooperation, new clean energy resources have become the priority of science and technology cooperation between China and Brazil and between

China and South Africa, and similarly, climate change topics are given priority in science & technology cooperative projects between China and India. China and South Africa carried out technological exchanges on coal liquidation and South Africa has introduced plasma ignition technology in its coal-fired boilers. China and India have conducted cooperative research exploring comparative energy systems and have conducted joint experiments on the climate and environment of the Qinghai-Tibetan Plateau. China and Brazil cooperated on hydropower projects. Climate change, as well as clean and renewable energy, was given priority attention in the short- and medium-term cooperation plan detailed in the New Delhi Action Agenda in 2004. The South-South cooperation against climate change among the BASIC countries functions on the principle of mutual benefit, yielding win-win results in economically, technologically, and politically.

B *Current Status of South-South Cooperation against Climate Change in China*

China engages in South-South cooperation by offering fully constructed equipment, joint technical projects, education, training, etc. South-South cooperation in the field of science and technology includes joint research, resource investigation, training, demonstrations, construction of research institutions, and donations of instruments and equipment.

The national foreign-aid budget in 2012 amounted to RMB 14.4 billion, and a large proportion of this aid has a direct or indirect relation with climate change. China addresses climate change in South-South cooperation primarily by engaging in technical training, disaster monitoring, provision of clean energy equipment, and construction of related infrastructure. For example, China presented North Korea, Mongolia, and other countries with PC-VSAT meteorological data satellite radio reception systems in accordance with the Voluntary Cooperation Plan (VCP) of the World Meteorological Organization (WMO), enabling them to receive meteorological data exchanged by the Chinese Global Telecommunication System (GTS) and Numerical Weather Prediction (NWP). China has also offered assistance in the form of meteorological instruments to over 70 developing countries and has helped over 100 countries train directors and scientific and technological personnel for their meteorological bureaus. Additionally, China presented more than 20 countries, such as Bangladesh and Indonesia, with the GeoNetcast DVB-S system, which can receive EOS/MODIS data, remotely-sensed data from meteorological satellites such as the FY-2 series, FY-1 series, and NOAA series. According to the China-Africa science and technology partnership plan, China popularized key technologies used for drought-enduring and high-yield wheat and corn, increasing grain output in various African countries such as Egypt, Sudan, and Mozambique. The remarkably enhanced capability of local crops to adapt to climate change has reduced stress surrounding grain security. China rolled out a drip irrigation project which helped spread the technology throughout greenhouses in Central Asia and improved the water saving capabilities of Africa, certain Arab areas, and the Caribbean. China also carried out a pilot project exploring rainwater utilization in African countries such as Kenya, which effectively eliminated water shortages in the local areas where it was applied. China promoted renewable energy and energy conservation as well, pushing the application of solar energy, small hydropower generation stations, and biogas in developing counties, which boosted local economic development and reduced poverty. For instance, China helped Tanzania craft a five-year development plan on national solar energy, it helped the Ivory Coast make a plan promoting photoelectric production, it popularized solar energy irrigation systems and water heaters in Zimbabwe, it invested in the construction of micro-hydropower stations in many African countries, it assisted Micronesia in biogas establishing demonstration projects, and it set up a high-efficiency off-grid lighting center and conducted semiconductor lighting demonstrations in Kenya. China cooperates with international organizations such as the UN Industrial Development Organization (UNIDO), the World Bank, and United Nations Environment Programme (UNEP) to provide joint aid projects. For example, China cooperated with UNIDO to conduct technological research on energy-saving building materials and to popularize the findings in a large number of developing countries. It participated in tropical disease research and related training sessions organized by the World Health Organization. And it offered tropical disease medicine and vaccines to developing countries and improved their diagnosis and treatment capabilities.

The examples listed above were predominantly reliant on intergovernmental cooperation mechanisms, with only a few based on the support of enterprises or non-governmental organizations. The projects focus on the areas that cause the greatest concern to recipient countries, such as natural disasters management, water access, clean energy production, and ecological environment. China's experiences and technologies can serve as demonstration and provide greater incentive and motivation, providing a more practical and target South-South cooperative relationship.

III South-South Cooperation Experiences and Obstacles

Great progress is being made in expanding new arenas of South-South cooperation as new space and opportunities arise alongside advancing globalization.

An overall increase of the power held by developing countries and 50 years of experience engaging in cooperative projects provide advantages in efforts to expand cooperative efforts to new topics. The cooperation extends to technology, education, information, and human resources, incorporating both domestic economy and trade. South-South cooperation also promotes the establishment of a new fair and reasonable international order that provides conditions favorable to international economic development and engenders an environment of "cultural communication" and "civilized dialogue." A cooperative international security mechanism is being established so that conflicts related to developing countries will draw adequate attention from the international community and so that developing counties can have a larger say in international security. South-South cooperation against climate change has also become an important topic within the international South-South cooperation mechanism. The mechanism allows developing countries to connect issues related to climate change with development of state power and plans for domestic sustainable development.

China's climate change targeted cooperation with Africa has consisted primarily of aid for agricultural water conservation projects, disaster prevention, other relief projects, national and quality of life infrastructure projects, and direct financial and labor aid, all given without any preconditions. The actions have deepened and strengthened an important alliance traditionally held between China and Africa and has led to each to better understand the other's perspective, and hence support and trust each other during climate talks. China is also training officers to be familiar with its national conditions and training senior professionals with a specialty in fundamental science and technology in South-South cooperation. These talents are necessary to manage the social development and the growth of local economies. Personnel with knowledge of African culture and national conditions are appearing in China as well, promoting further commercial, trade, cultural, and technological communication between China and African countries. China shares more common interests in cooperation with some other emerging developing countries. The cooperation further promotes research, development, popularization and application of science and technology that addresses climate change, as well as industrialization of scientific and technological achievements that address climate change. The cooperation enhances mutual understanding and support, provide a greater common ground that leads states to adopt the same positions on some key issues in climate negotiations, and thus better protects the rights and interests of developing countries.

It is important to note that technology cooperation and transfer between developed countries accounts for over 80% of all trade, while that between a developed country and a developing country and among developing countries accounts for less than 20%. The cooperative gestures among developing countries are even smaller, especially with regard to climate change. Although some developed countries reduced economic aid to developing countries due to the global financial crisis in recent years, many developed countries still took active measures to support developing countries with technologies addressing climate change. In 2009, the European Union initiated a program of technological partnership with Africa and put in an initial 65 million euros to support projects related to climate change, clean water, and energy. Germany planned to carry out research on climate change, water resources management, and desertification and land management in sub-Saharan Africa and set up a "regional capacity center." In 2009, the UK launched Phase III of a project that function as a cooperation with China to develop climate change adaptation strategies in its earlier phases, carrying out scientific research, policy research, and experiments in the areas of climate change, drought, urban development, and human health. 2009 was also the year that the United States proposed the establishment of a technological partnership with Muslim countries and sent scientific emissaries to scout out these cooperative opportunities in the fields of climate change, energy, and green and low-carbon technologies. Japan initiated a sustainable development oriented scientific and technological research partnership program and implemented scientific and technological aid plans in Africa, centering on environmental protection, disaster response, energy, and grain production. Developed countries conduct technological cooperation and transfer by executing bilateral cooperative agreements, initiating various international plans, and guiding international organizations. The cooperation efforts can generally be divided based on how cooperation and transfer are undertaken and what motivates the agreements. Cooperation can be achieved through policy formulation and through cooperation of international or organizational associations and enterprises. Countries are motivated to engage in cooperation for political needs, market needs, and public welfare. As opposed to North-South cooperation, South-South cooperative structures must still strengthen their overall system plan, learn to emphasize the key points of interest, develop support systems for project implementation, and generally promote a greater cooperative environment. It is not easy to develop and transfer applicable climate change technologies that are environmentally friendly, affordable, practical, easily maintainable, and in line with local needs under the current South-South cooperative regime. Such transfer suffers from the following problems.

First, no overarching program that can adequately manage these transfers exists. Present South-South cooperation consists mainly of small scale deals that do not include adequate planning. The cooperative countries fail to find out their partners' real needs and do not fully understand the economic conditions, technical absorptive capacity, management level, and social environment of each partner faces, as well as applicability and readiness of the technologies being exported. As a result, some efforts remains at the planning stage and struggle to reach effective implementation or obtain the necessary recognition and promotion from partner countries. Many countries have also struggled to differentiate between government projects and market based private projects, and therefore have been unable to develop appropriate accompanying policy.

Second, there are no smooth cooperative channels and insufficient funds. Developing countries have a comparatively low economic level by definition, so channels for funds transfers related to South-South cooperative actions are restricted. Partner countries also habitually fail to utilize the market mechanism to address the issue. South-South cooperation is primarily conducted in the form of governments of large, emerging developing countries providing aid, cooperative research, and training. Many enterprises have been exploring ways to become involved but most developing countries lack the necessary channels to acquire information and applicable technologies, and intermediaries serving South-South technological cooperation do not have highly developed coordinating capabilities and resource integration awareness.

Third, cooperative efforts are not properly popularized. Developing countries do not adequately publicize the achievements made in South-South cooperation and western media rarely report on them. The cooperative countries do not engage in adequate follow up of the experiences and lessons learned from the projects undertaken, and thus do not initiate other similar projects.

Despite the great progress made by China in forging forth with South-South cooperation, it has also run into the problems mentioned above. Due to the diversified management and investment of South-South cooperation against climate change, China cannot exploit all of the possible advantages available through South-South cooperation.

IV Prospect for South-South Cooperation against Climate Change

Although South-South cooperation was originally intended for development and poverty relief, it has also attached great importance to global environmental problems such as climate change from the very beginning. Developing countries are industrializing at an ever more rapid rate. If they follow the example developed countries set in the 18th century, the global environment will be completely destroyed. Climate change raises urgent concerns for developing countries. For instance, a group led by Peru which included Chile, Columbia, Ecuador, Panama, and several other countries held a discussion in 2000 in an attempt to better understand and address the El Niño phenomenon. The *Mauritius Declaration* and *Mauritius Strategy* published in 2005 the further the implementation of the *Barbados Programme of Action for the Sustainable Development of Small Island Developing States* directly addresses the issue of natural disasters caused by sea level rise.

A South-South Cooperation against Climate Change has Shifted Values in a New Era

Developing countries share similar natural and social factors. By 1999, the world population was 6 billion. It is predicted that in 2011, the figure will increase to 7 billion, out of which nearly 6 billion (5.6 billion in 2008) will be living in developing countries. According to a survey conducted by the United Nations, the world population will increase to 10.1 billion by 2100. Some 97% of the 2 or 3 billion people that will join the world in the next four decades are expected to be born into developing countries, and nearly half them will be born in Africa. Developing countries cannot sideline climate change as a problem that is common to the whole world. They can neither ignore human influence on the climate nor climate change's influence on human survival.

Second, developing countries focus on shared interests. In current international climate change negotiations, developing countries are the greatest victims of climate change and form one half of major camps division (developed countries and developing countries) present in climate change negotiations. Though developing countries face different national conditions and find themselves at different levels of development and under dealing with different challenges, they have certain shared interests. Their attitude undoubtedly plays a pivotal role affecting the result of negotiations.

Third, climate change and development influence each other directly. Poverty and problems of basic survival in developing countries such as starvation, polluted drinking water, and sea level rise severely threaten the life of billions of people. These problems should be addresses conjointly with other problems such as emissions reduction demanded to hinder climate change. Developed countries should recognize a continuation of the "common but different responsibility" principle and engage efforts to help developing countries cope with urgent survival needs and to further sustainable development. Cooperation, harmonization, and balance among developing countries during negotiations can also provide references examples for each country to learn from.

According to the *National Communication* submitted to the Secretariat of UNFCCC by developing countries and the *Climate Technology Needs Assessment* jointly published by the Secretariat of UNFCCC and the UNDP. It is evident that many developing countries have an urgent demand for technology to aid in water resource utilization, agricultural technology, and renewable energy technology such as solar, biogas, and hydroelectric. They are also in need of disaster prevention and disaster reduction technology such as climatic system observations and predictions, and adaption to climate change in the fields of health, industry, architecture, transportation, capacity increases, and cultivation of public awareness. Certain developing countries especially those emerging countries with abundant technological and practical experience, will have a greater influence on other developing countries, and will drive forward South-South cooperation efforts. Pluralistic communication and cooperation among developing countries at international climate negotiations powerfully protects the interests of most developing countries and promotes the establishment and implementation of beneficial international conventions. Though the international situation is ever changing, a new international climate order has been finally established, but remains in need of solidarity, mutual understanding, and cooperation among developing countries.

B *Challenges Ahead*

New situations have undoubtedly raised new challenges for South-South cooperation to properly address climate change within international discussion. These challenges are caused by the following reasons.

1 Developing Countries have Shifting Interests

As evidenced in the climate change negotiations in Copenhagen, Cancun, and Durban, although developing countries and developed countries continue to form two mainly opposing camps, developing countries are no longer as firmly committed to each other, and internal disagreement of varying severity has emerged among developing countries, centering on differences in development stage, historical and political background, and level of interest. The international long-term objectives and the determination on emissions peak centered on a "shared vision," a dual-track system related to the climate system, the applicable scope of financial support, channels for financing, and targets receiving support all suffer contention among the countries. Developed countries are setting up new international rules that threaten developing countries differently depending on their respective situation, creating varied requirements for carbon emissions, forming international carbon markets, and carbon trade, and pushing other projects that ultimately make it more difficult for developing countries arrive at complete agreement.

2 South-South Cooperation Mechanisms must be more Innovative

South-South cooperation against climate change relies almost exclusively on traditional cooperation styles and centers on improving adaptation to climate change and shifting development in developing countries onto a low-carbon track. Developing countries must strive for greater independence and use South-South cooperation to find innovative win-win solutions. Current cooperation relies heavily on governments, but does not engage nearly enough in government guidance and support of free market forces, which is necessary to better improve project financing and technological innovation.

South-South cooperation currently finds itself with both great opportunities for further development but also great challenges to realize those opportunities. It is imperative that a better conceptual consensus is developed through active communication so that agreements can have a greater impact. We should respect our planet and not follow example of developed countries in order to create a more harmonious international community and an ecologically vibrant earth as well as to eradicate poverty and improve the overall standard of living. This should be the unifying theme of South-South cooperation when it convenes for international political and economic decisions. Only this approach will allow developing countries to help each other promote the coexistence of humans and the natural environment.

C *Further Thoughts on International South-South Cooperation against Climate Change*

A grand, complicated economic and political background unfolds behind climate change. In order for South-South cooperation to effectively combat climate change quickly, its primary world players, including China, should pay special attention to some ideas and popularize them as necessary, and of course, they must take some actual steps to combat climate change themselves. The concepts of substantial effects, diversified structures, mutual benefits, and common development are keys in this struggle. Here are some actual steps to undertake.

1 A Special Fund for South-South Cooperative Efforts against Climate Change should be Established

A cooperative fund for developing countries should be established, to be used primarily to improve recipient countries resiliency and independence, especially in the fields of disaster prevention and relief, climate change adaptation, capacity expansion, and clean energy development and utilization. The fund can also be used for promoting communication and cooperation, honing cooperation among developing countries against climate change, promoting applicable technology transfer related to climate change adaption and mitigation, and improving developing countries capacity to address climate change.

2 Encourage Developing Countries Technology Exports through Various Channels

In addition to governmental support, South-South cooperation should guide and encourage enterprises and non-governmental organizations to play a greater role in South-South cooperation against climate change, and should lead quickly evolving developing country enterprises to greater export potential.

3 Strengthening Coordination Mechanisms

South-South cooperation should make good use of existing coordination and cooperation mechanisms and establish more such mechanisms among developing countries in various fields and at various levels. These will aid in deliberating and determining the strategies, policies, and major cooperative projects against climate change. These developments will produce an effective communication mechanism and enhance the composite force of South-South cooperation.

(This article was originally published in Chinese in 2012.)

CHAPTER 3

China's Sustainable Development in a Shifting Global Context

Wang Yi

The international community reached a consensus on sustainable development at the United Nations Conference on Environment and Development (UNCED) in 1992, but since then there have been substantial changes to the global environment. Sustainable development has advanced worldwide and there have been important strides in addressing the Millennium Development Goals (MDGs), especially as China and other emerging economies become stronger and alter patterns of global political economy, resources, and environmental security, creating ever closer links between China and the rest of the world. But the world's population is rapidly increasing, poverty remains rampant, global warming is at the center of many international discussions, there are severe environmental pollution problems across various regions, the imbalance between supply and demand for key resources and energy is increasing, and environmental and developmental concerns face new dilemmas in trying to maintain fairness and equality. There are great challenges ahead in the efforts to achieve sustainable development.

A green economy is emerging despite of the daunting impact of the global financial crisis. "The green economy in the context of sustainable development and poverty eradication" had been established as one of the two themes for the UN Conference for Sustainable Development (Rio+20) held in Rio de Janeiro, Brazil in June, 2012. It aims to further eliminate poverty, change unsustainable production and consumption patterns, and protect and manage the natural resources foundation of economic and social development, by promoting a green economy and an institutional framework for sustainable development. To ensure that that the strategy is effective, country specific, oriented toward long-term success, and able to capitalize on opportunities, it is essential to reevaluate the structure of sustainable development projects, identify gaps,

* Dr. Wang Yi is a Professor and Deputy Director-General of the Institute of Policy and Management (IPM), Chinese Academy of Sciences (CAS). He is also the team leader and chief scientist of the annual *China Sustainable Development Report* (CSDR).

 | DOI 10.1163/9789004274648_004

and summarize experiences and lessons learned with a global perspective. Such a strategy will facilitate joint efforts to shape a sustainable future that will be resource-efficient, environmentally-friendly, green, low-carbon, fair, inclusive, and more competitive.

I The Global Process of Sustainable Development: Transition, Opportunity, and Uncertainty

The concept of sustainable development is continually evolving. Its definition in *Our Common Future* is "development that meets the needs of the present without compromising the ability of future generations to meet their own needs."[1] A sustainable development strategy broadly aims to promote harmony between human beings and nature, and more specifically it implicates natural sustainability, i.e. inter-generational justice in the distribution of resources and intra-generational justice between regions. The more philosophical view of sustainable development has triggered a widespread debate focused on the difficulty of implementation. This strain of sustainable development is particularly difficult to define and explain, and there is no agreement on how it should be measured.

Despite these difficulties, the international community embraced sustainable development as a strategic framework at UNCED in 1992. It was further expanded upon at the World Summit on Sustainable Development (WSSD) held in Johannesburg, South Africa in 2002 as encompassing three interdependent pillars: economic development, social development, and environmental protection. Although many countries have developed sustainable strategies and national action plans, making some progress in certain sectors, the extension and generalization of this relatively abstract concept that lacks particular details has turned it into a mere political slogan in many instances, used or distorted by interest groups to pursue personal interests or objectives. The absence of legally binding multilateral agreements and compulsory domestic implementation mechanisms allow for diversion, distracting from the essence and authentic objectives of sustainable development.[2]

1 WCED. 1987. *Our Common Future*. Oxford: Oxford University Press.

2 Victor, D.G. 2006. Recovering Sustainable Development. *Foreign Affairs*, 85(1): 91~103; CAS Sustainable Development Strategy Study Group 中国科学院可持续发展战略研究组. 2008. *China Sustainable Development Report 2008—Policy Review and Outlook* 2008 中国可持续发展战略报告——政策回顾与展望. Beijing: Science Press.

Fortunately, since the 2008 global financial crisis, developing a green economy is becoming a popular option for addressing the multiplicity of challenges facing each country, and it offers a new opportunity for realizing global sustainable development and for achieving green transformation.[3] Under active advocacy by UNEP, OECD, and other international organizations, green development strategies have been formulated in many countries and regions. For example, the European Commission has presented roadmaps for developing a low-carbon economy, establishing greater resource efficiency e, and constructing a secure, competitive, and carbon-free energy system;[4] the Korean government promulgated the "Framework Act on Low Carbon Green Growth"; and China pushed green and low-carbon development in its "Outline of the 12th Five-Year Plan (2011–2015) on National Economic and Social Development."

A green economy is another concept that has been around for a while and bears similar connotations to sustainable development. Most discussion of the green economy is currently surrounding the financial crisis, with some international organizations and experts calling to address the crisis and climate change simultaneously by implementing "green governance" and developing the green economy, which would further advance multilateral negotiations on sustainability.[5] Although the green economy also lacks a universally accepted definition, it is widely understood as aiming to develop green emerging industries, increasing green jobs, and enhancing competitiveness that is environmentally friendly, thus promoting a greening of the global economy and building toward sustainable development.

3 CAS Sustainable Development Strategy Study Group 中国科学院可持续发展战略研究组. 2010. *China Sustainable Development Report 2010—Green Development and Innovation 2010* 中国可持续发展战略报告——绿色发展与创新. Beijing: Science Press; CAS Sustainable Development Strategy Study Group 中国科学院可持续发展战略研究组. 2011. China Sustainable Development Report 2011—Greening the Economic Transformation 2011 中国可持续发展战略报告——实现绿色的经济转型. Beijing: Science Press.

4 European Commission. 2011a. *A Roadmap for moving to a competitive low-carbon economy in 2050*. COM (2011)112. Brussels; European Commission. 2011b. *Roadmap to Resource Efficient Europe*. COM (2011)571. Brussels; European Commission. 2011c. Energy Roadmap 2050. COM (2011)885. Brussels.

5 UNEP. 2008-10-22. *"Global Green New Deal"—Environmentally-Focused Investment Historic Opportunity for 21st Century Prosperity and Job Generation*. UNEP http://www.unep.org/documents.multilingual/default.asp?documentid=548&articleid=5957&l=en; Edenhofer O., et al. 2009. *Towards a Global Green Recovery: Recommendations for Immediate G20 Action. Report Submitted to the G20 London Summit*. Report on Behalf of German Foreign Office.

While negotiations on green economy projects and the "Rio+20" conference are important, expectations of the international community are not high. The 2009 Copenhagen climate conference failed to reach an agreement. Even though the next climate conference in Durban in 2011 approved a transitional package, the prospects for achieving a long-term carbon reduction framework by 2015 remain optimistic. Meanwhile the difficulties faced in recovering from the financial crisis have reduced people's attention to climate change, so that there is less enthusiasm around a green low-carbon economy. The lack of strong and effective leadership opens the door for wait-and-see strategies, and financial pragmatism prevails. Countries are reevaluating their strategies and taking more pragmatic, comprehensive, and long-term countermeasures. Serious disagreement persist around such issues as funding, technology, and equity, requiring major efforts and clever deals if any substantive progress beyond Rio+20 is to be achieved.[6] Similar to the climate negotiations, international negotiations around the theme of "green economy" will encounter the following key issues and conflicts:

- First, responsibility, equity, and political commitment will be hotly debated. The "common but differentiated responsibilities" principle established 20 years ago remains at the center of any discussion. Because of the prolonged debate on this principle, funding and technology transfer commitments made by developed countries to support sustainable development and to address climate change in developing countries have not been implemented, making it difficult for many developing countries to meet MDGs and leaving them without assurance of the right to develop. Green economy debates also raise the issue of equity. Without an explicit political commitment by developed countries, it is impossible to develop common goals, time frames, and roadmaps for the green economy. Increasingly, there are loud voices calling on China to undertake greater international obligations and there is increasing pressure on China to reduce its emissions, showing the increasingly diversified interests in the field.
- Second, there exists a divide over the understanding of a green economy and its appropriate definition. Developed countries hold the reduction of carbon emissions as the central goal of the green economy because they

6 Kuper, S. 2011-09-17. *Climate change: who cares any more?* FT Magazine. http://www.ft.com/cms/s/2/1b5e1776-df23-11e0-9af3-00144feabdc0.html#axzz1mYWgzGcL; Wang Yi 王毅. 2011. 学做大国从 "绿色" 开始 To be a great power starting from "greening." 财经年刊 "2012: 预测与战略" *Caijing Annual Edition—2012: Forecasts and Strategies*, 290–293.

have solved most conventional environmental issues. It is more urgent for developing countries to improve resource efficiency and address conventional pollution concerns and environmental issues due to the breadth issues which they face. A diversity of goals, technologies, paths, and road maps should be embraced in developing the green economy in different countries, depending on the respective stage of development, national conditions, regional environment, and international responsibilities. Goals cannot themselves solve the issues at hand; rather an integrated solution is needed.

- Third, many conflicts exists surrounding green trade. In light of the impact of the financial crisis, the enhanced competitiveness of emerging economies, and the lack of legally binding power of multilateral environmental agreements (MEAs), it is unsurprising that some countries, especially developed countries, have established "green trade barriers" and "IPR barriers," employing such measures as tariffs, WTO trade rules, IPR protections, and trade protectionist legislation. These measures negatively affect trade in environmental goods and services and the diffusion of green and sustainable technologies. The EU's decision to extend the reach of its Emission Trading Scheme (ETS) to foreign airlines, "301" investigations and "anti-dumping and anti-subsidy" investigations by the US against Chinese clean energy enterprises, and litigation by the US and some European countries against China for its export controls on some raw materials are just the first conflicts in the "green trade war," which will certainly become increasingly severe.

There will be various new challenges in the process of developing a green economy. Over the last two decades, about two to three billion people will enter the industrialization phase of heavy manufacturing and chemical industry development, and they may face legally binding constraints on carbon emissions in the form of caps and demands for quantitative emissions reductions. The challenge for emerging economies, and the world as a whole, will be to appropriately grasp development opportunities, share and utilize limited resources, energy, and emission space in a rational and equitable manner in order to accumulate wealth and modernize, as well as gaining a competitive advantage in the field. A cooperative framework for the development of a global green economy should be established collaboratively, and it must encompass concrete actions, just distribution of obligations, the right to develop, and the sharing of best practices. Various conflicts and pressures will be alleviated by improving resource efficiency, developing clean energy, investing in green innovation, and changing unsustainable development and trade patterns, realizing a common human vision of sustainable development.

II China's Contributions to Sustainable Development: Practice, Experience, and Global Challenges

Over the past two decades, China's sustainable development practices have enriched the concept of sustainable development internationally and have contributed to global sustainable development efforts.

China engaged in the discussion and drafting of the report on "Our Common Future," leading it to become one of the first countries to formulate and implement a sustainable development strategy. The Chinese government defined environmental protection as fundamental national policy in 1982 and signed the "Rio Declaration on Environment and Development" and the "Agenda 21" in 1992. In 1994, China took the lead in issuing a national formulation of Agenda 21—"China's Agenda 21—White Paper on China's Population, Environment, and Development in the 21st Century." In 1996, sustainable development was officially designated as fundamental national development strategy, thus transitioning sustainable development from scientific discussion to government mandates and concrete actions, including institution building, policy development, organizational management, resource conservation, environmental protection projects, as well as green and low-carbon pilot programs. The predominant experiences and contributions China has contributed to innovation of global sustainable development systems include theoretical innovations, system creation for comprehensive plans and targets, holistic institutional arrangement, policy and measure adoption, project implementation, and mass application. The following are examples of these contributions.

- China has presented a series of sustainable development ideas. As China entered the 21st century, assented into the WTO, and entered into a development stage characterized by heavy industrial growth and chemical industry expansion, as well as rapid urbanization, it has become the "factory of the world" and the second largest economy in the world. Tense realities regarding resources, energy, and environment became more frequent throughout China in 2002. To address the issues of environment and development, the Chinese government has presented a series of new sustainable development idea and has taken concrete actions to implement these ideas, creating a Chinese style of sustainable development. These new ideas include: the new path to industrialization (2002), a scientific outlook on development (2003), the circular economy (2004), a resource-efficient and environmentally-friendly society (2004), the harmonious society (2005), an innovative country (2006), an ecological civilization (2007), a green economy and a low-carbon economy (2009), the transformation of economic development patterns (2010), and green and low-carbon development (2011). Some of these ideas have been formulated and developed based on China's own experience and knowledge, some are based on international experience, and others have been formulated alongside other countries, occasionally with China heading the effort. China has formulated a sustainable development strategy and comprehensive plan with legally binding "energy conservation and emissions reduction" targets at its core. For the period of the "11th Five-Year Plan (2006–2010)," the Chinese government set binding targets of 20% increases in energy efficiency and a 10% reduction in emissions for major pollutants. It also developed corresponding work schemes and key programs which were implemented through integrated legal, administrative, economic, and technological measures. In 2009, China made a voluntary commitment to reduce its carbon intensity and increase its forest carbon sink, thereby addressing climate change in its energy conservation strategy. During the period of "12th Five-Year Plan," following the policy trend set by the "11th Five-Year Plan," the Chinese government proposed changing the pattern of economic development as a primary objective, increasing the proportion of non-fossil fuel energy sources and other meeting other binding targets, as well as formulating new policies to rationally control total energy consumption. The goal is to gradually establish a carbon emissions trading market in order to promote China's transformation toward green and low-carbon development, opening a path toward sustainable development ideas and practices with Chinese characteristics.

- To implement its sustainable development strategy and in order to meet energy conservation and emissions reduction targets, the Chinese government has created a series of legal edits and additions, including the Cleaner Production Promotion Law (2002), the Environmental Impact Assessment Law (2002), the Water Law (2002), the Renewable Energy Law (2005), the Circular Economy Promotion Law (2008), and amendments to the Energy Conservation Law (2007) and the Water Pollution Prevention and Control Law (2008). China has also launched the National Program on Climate Change (2007), established a national group to lead efforts on climate change, energy conservation, and emissions reduction, and designated a special management body for climate change (2008). The National People's Congress approved the "Decision on Actively Addressing Climate Change," which provided the legal foundation for implementing of the measures listed.
- China has implemented large-scale ecological protection projects, energy conservation, and environmental protection projects. Since 1998, China has made large-scale investments in ecological conservation programs and environmental protection infrastructure. The funding for ecological

conservation in 1998 exceeded that of the two previous decades combined. In the "10th Five-Year Plan (2001–2005)" period alone, China put 700 billion RMB into six key forestry programs including the "natural forest protection" and "grain for green" projects (Deng Huaning, et al., 2005.) These projects have been running for over ten years and have demonstrated significant achievements. Since 2004, the Chinese government has launched one engineering project after another, all aimed at the concepts of circular economy, resource conservation, and renewable energy. Since 2008, energy conservation, emissions reduction, and ecological environmental construction have been specified as primary items in the economic stimulus plan, greatly improving China's environmental infrastructure.

China made substantial achievements in energy conservation and emissions reduction during the"11th Five-Year Plan" period thanks to the efforts mentioned above. Energy consumption per unit GDP decreased by 19.1%, and the total amount of COD discharge and SO_2 emissions declined by 12.5% and 14.3% respectively. Wide spread use of renewable energy became common place. By the end of 2010, China's grid-connected wind power capacity amounted to about 30 million kW, an average annual growth of 94.5%,[7] making it the world leader in installed wind power capacity. China has also developed and installed some energy-saving and emissions reducing technologies and equipment in at a level ranking high in international comparisons (e.g., clean coal power generation).

The success of China's energy conservation and emissions reduction strategy is reflected in ways beyond the figures listed above. Comprehensive outcomes have significantly raised the government officials' awareness at all levels, and the public is likewise better informed of energy conservation and emissions reduction efforts. Visible effects of China's new approach include taking concrete actions in priority areas, "learning by doing," implementing multisectoral and multidimensional pilot activities (including circular economy initiatives, ecological industrial parks, low-carbon projects, and sustainable development efforts in pilot regions,) accelerating the phasing out of outdated production machinery, cleaning the industrial sector, greatly improving innovative and technological capacity, and substantially increasing the size of green and low-carbon related industries. China may reach advanced world levels of energy conservation and emissions reduction technologies, equipment

7 State Electricity Regulatory Commission 国家电力监管委员会. 2011-08. *2010 Annual Business Briefing of Electricity Generation* 2010 年度发电业务情况通报 http://www.serc.gov.cn/ywdd/201109/W020110901610165944272.doc.

manufacturing, and project construction and management in major industries by 2020 or 2030 if energy conservation and emission reductions policies are adhered to over the next decade.

There is no doubt that there are still some problems and a need for further improvement experience with China's sustainable development, energy saving, and emissions reduction strategies. The outstanding issues include: a top-down decision making process and over-dependence on administrative measures which have resulted in an inefficient use of resources and sector specific and other special interest groups which have hindered reform expansion, jeopardized national interests, and wasted governmental efforts due to the lack of effective coordination mechanisms. As China grows stronger, the relationship between China and the world has drastically changed. Almost all of China's resource and environmental issues are of global significance because of their proportionally large scale effects, for which neither China nor the world are fully prepared.

China's increasing strength has led its relationship with the rest of the world to become increasingly interactive, such that no action by China goes unnoticed. It has been more than 10 years since China ascended to the WTO and implemented its "going out" strategy. It is relevant to note that China's rapid economic growth has paralleled its engagement in the process of globalization. Given its resource endowment, the size of its economy, and its status as "the world's factory," China is increasingly dependent on foreign sources for key resources such as oil, iron ore, copper, and potash. Over 50% of China's oil and iron ore is imported from foreign countries. China's carbon emissions account for more than 20% of the world total and continue to increase. As demand for foreign resources increases and overseas investment grows, two prominent opinions prevail: some believe that China should take the world lead in the green economy by taking advantage of the momentum already present in energy conservation and emissions reduction, while others say that China is implementing "neocolonialism" and are curious and concerned about whether China's great power will change the existing global rules of the game. Some Chinese enterprises engaging in business activities overseas are unfamiliar with international practices, do not attach great importance to their social and environmental responsibilities, and do not respect to local culture sufficiently, causing damage to local resources and the environment, and leading to social conflict. Beyond losing assets, these companies also jeopardize China's national interests, creating a negative perception. In light of this concern, the Chinese government has formulated some regulations in an attempt to reign in the environmental and social conduct of those enterprises and to avoid risks. The current effort does not go nearly far enough and requires attention from higher levels and additional sectors if effective measures and strategies are to be implemented.[8]

The green economy has become a popular theme globally. Developing a green economy and promoting sustainable development do not hinge on the acts of a single country, but rather on the broader principles of equity, competitiveness, and cooperation. To become a responsible country, China should start learning to promote "green globalization," and should push for inclusiveness and responsibility in all dimensions for both oneself and others.[9] Achieving a "green rise" requires a better integrated strategy, a historical and global vision, and the assumption of additional international obligations that are compatible with the expanding capability. There remains a long road ahead.

III Path to the Future: Institutional Reform, Policy Development, International Justice, and Comprehensive Transformations

Although some debate and uncertainty surrounding the future details of a green economy and low-carbon development remain, the general direction is clear. China must pursue opportunities, create an empowering environment, control the pace of development, enhance innovation, take initiatives, seek cooperative win-win solutions, and even take the lead on green initiatives when the situation calls for it. There are two aspects which must be considered while following a sound and progressive green development path in order to gradually realize a comprehensive green transformation. One, in light of the upcoming carbon cap, strategic choices for the next decade are very important. A balance between making use of traditional advantages and pursuing green innovations must be maintained. Two, in order to secure national interests and peaceful growth, China must actively participate in the process of restructuring global environment and development, and create strategic global areas through cooperative arrangements.

8 State Forestry Administration and Ministry of Commerce 国家林业局、商务部. 2009-03-26. *Guide on Sustainable Overseas Forests Management and Utilization by Chinese Enterprises* 中国企业境外森林可持续经营利用指南; China Banking Regulatory Commission 中国银监会. 2012-02-24. 关于印发绿色信贷指引的通知 *Guideline on Green Credit. CBRC issued document* [2012] *No. 4* http://www.cbrc.gov.cn/chinese/home/docView/127DE230BC31468B9329EFB01AF78BD4.html.

9 Wang Yi 王毅. 2011. 学做大国从"绿色"开始 To be a great power starting from "greening." 财经年刊 "2012: 预测与战略" *Caijing Annual Edition—2012: Forecasts and Strategies*, 290–293.

A *Formulating China's Global Resource and Environmental Security Strategy*

The following steps are important for China's strategy in securing global resources and ensuring environmental security: 1) China should clarify its national interests and strategic goals in the resource and environmental sectors by participating as stakeholders and increase the transparency of resource and energy demand, effectively reducing the risks caused by information asymmetry, and protecting Chinese interests by political, diplomatic, economic, technological, and military means. 2) China should strengthen education and research, improve the capacity to assess development trends in resources and energy, learn more about the dynamics of import and export of key resources and energy, and propose workable policy recommendations. 3) Beginning with its neighbors, China should develop long-term and stable cooperative relationships with foreign countries by implementing long-term cooperative projects such as the construction of infrastructure related to resource exploitation, technological assistance, and capacity building. 4) China should strategically acquire resource and energy-related products and technologies, simplify approval procedures for such acquisitions, and encourage state-owned and private enterprises working with these resources and technologies to engage in equity and technology mergers.

B *Promoting Structural and Process Innovations in the Field of Sustainable Development*

Serious fragmentation in the field of sustainable development in China hampers coordination between interest groups and is not conducive to promoting preexisting advantages and enhancing efficiency. It is suggested that the macro-ministry reform its framework to reorganize and establish the following bodies: 1) a Ministry of Energy and Climate Change to integrate existing government functions related to energy development, energy conservation, climate change, and low-carbon development, unifying the tasks of planning, policy making, supervision, and management in these closely linked fields; 2) a Ministry of Resources and Environment, to integrate governmental functions related to resources and environmental components. The merging of water, forestry, and environmental protection departments should be implemented first in order to effectively address issues surrounding development regionally and along river basins; and 3) an International Development Agency that would play an important role in coordinating between multiple government departments for implementation of the "going out" strategy, development of international assistance plans, enhancement of the state's "soft power," securing the resource-environment, and advancing international environmental and development order. It is imperative that a large country with increasing responsibilities establish an independent agency for this task.

C *Grasping the Driving Forces Underlying Sustainable Development Policies*

China should adjust its population policy to support sustainable economic growth, alleviate the negative impact of an aging population, and provide a new labor force. It should place priority during the next 10 years on improvements in energy efficiency and benchmark management. Policies should attempt to facilitate the "decoupling" of economic growth from resource consumption and negative environmental impacts. Resource and energy price reform should take precedence over other financial and tax policy reforms while reform of land distribution and management should be accelerated. Emphasis should be placed on developing an integrated scheme for addressing environmental issues that cross administrative regions and river basins. Energy conservation and low-carbon policies should be adhered to, carbon emissions trading markets should be constructed with caution by creating airtight macro structures and improving diversification capacity. For the markets to function it is also important to issue relevant guidelines and regulations and avoid blind leaps and unnecessary resource and capital waste.

D *Altering Foreign Economic Cooperation Strategies and Enhancing Corporate Social Responsibility for Overseas Development*

China's increasing economic strength has raised certain concerns in the international community, bringing about a need for changes to China's foreign economic cooperation strategy. There are three primary aspects to focus on: 1) China must develop a modern comprehensive foreign economic cooperation strategy, emphasizing energy conservation, emissions reduction, and climate change in guiding cooperative negotiations; accelerate the comprehensive transformation of foreign economic development patterns including diplomacy; facilitate the broad participation a coordination of stakeholders including diplomats, business officials, information technicians, departments and enterprises working on development, and the public. These efforts will more rationally develop and protect overseas resources and energy. 2) In the next round of WTO negotiations, China should propose a request for coordinating WTO rules with MEAS so as to avoid potential trade conflicts over GHG emissions, national environmental protection, green industry development, IPR protection, natural resources, and raw materials. 3) China should adjusting its "going out" strategy, thereby transforming itself to be more conducive to green investment and trade, and it should formulate guiding principles for Chinese enterprises engaging in overseas development. In addition complying with business rules and international best practice, these enterprises must strengthen their management, normalize their investment and development conduct, respond to their social and environmental responsibilities where their business activities are located, support local capacity-building for sustainable development, and promote the establishment of international rules for overseas corporate social responsibility and industrial transactions. 4) China should change its patterns of foreign aid, establishing green aid mechanisms, taking energy conservation, emissions reduction, and climate change as the focus of international aid, strengthening South-South cooperation, promoting the development and protection of overseas resources and energy by directly or indirectly using international assistance funds, and establishing a green image for the country and for enterprises.

China faces unprecedented challenges in domestic and international development and environment, with no fully developed experience or models to follow. Developing sustainably and development of a green economy requires engineering comprehensive systems with clear targets and an integration of legally binding regimes and incentive mechanisms, utilizing the combined effects of various policies, facilitating synergy between different environmental and development objectives, methods, costs, and pathways, and promoting systems innovation in institutional construction, supporting policies, technical research and development, capacity-building, business models, and market development, creating a modern ecological civilization. Taking the "Chinese path" is of great significance not only for China's development, but also for global sustainable development, especially if China's experiences can be used as a roadmap for the rest of the world.

(This article was originally published in Chinese in 2012.)

CHAPTER 4

Distribution and Implementation of Energy Saving and Low Carbon Development Targets for the 12th Five-Year Plan

Chen Ying

Abstract

With the widespread efforts across society, China reduced energy consumption per unit of GDP by 19.1% during the 11th Five-Year Plan (FYP) period, nearly hitting the 20% target. For the 12th FYP, China will continue to promote energy conservation and low-carbon development despite the many challenges that exist. This paper summarizes low-carbon targets and related policies and measures for the 12th FYP, focusing on regional target distribution and specific initiatives to achieve these targets.

Keywords

energy conservation – emissions reduction – the 12th Five-Year Plan (FYP) – target distribution

A set of binding, quantified energy conservation and emissions reduction targets were first issued in the 11th Five-Year Plan (FYP) of national socioeconomic development, for example to reduce energy consumption per unit of GDP below 2005 levels by 2010 and to reduce emissions of key pollutants, such as COD and SO_2 by 10% below 2005 levels. The 11th FYP also included the first consideration of greenhouse gases (GHGs) in a qualitative and objective manner for economic and social development planning. According to Announcement No. 9 released by the National Development and Reform

* Chen Ying, Senior Research Fellow, Institute for Urban and Environmental Studies (IUE), Chinese Academy of Social Sciences (CASS). Her interests include global environmental governance, energy and climate policy. The author thanks contributions from Dr. Zhou Yamin, IUE post-doctorate, CASS for translation and formatting.

 | DOI 10.1163/9789004274648_005

Commission (NDRC) and the National Bureau of Statistics of China in June 2011 (see Appendix VI), energy consumption per unit of GDP was reduced by 19.1%. All regions aside form Xinjiang, which uses a separate assessment, have achieved their targets; 28 regions succeeded beyond their set targets and 10 are excelling in particular in comparison with the rest. During the 11th FYP period, China's average annual energy consumption growth rate was 6.6%, much lower than the average annual economic growth rate of 11.2%. The energy consumption elasticity coefficient fell from 1.04 during the 10th FYP period to 0.59, reversing the upward trend of energy intensity that is common in the process of accelerated industrialization and urbanization, supporting both stable and rapid economic development, and making great contributions to combating climate change.

In March 2011, new energy-saving and emissions reduction targets were proposed in the 12th FYP of national socioeconomic development. There exist many challenges for all levels of government to find effective ways to achieve these new targets.

I Energy Saving and Emissions Reduction Targets for the 12th Five-Year Plan and the Challenges Ahead

The outline for the 12th FYP on national economic and social development of the People's Republic of China (hereinafter referred to as the Outline) was officially released on March 16, 2011. It emphasized new progress in scientific development and a modal transition of economic development over the next five years. Building a resource efficient and environmentally friendly society is key to accelerating this transition. The Outline also emphasized resources management and environmental protection as a principle national strategy along with pursuing sustainable development by saving energy, reducing greenhouse gas emissions intensity, promoting a circular economy, deploying low-carbon technologies, actively addressing global climate change, coordinating relationships between social and economic development and the general population, resources, and the environment.

A *Energy Conservation and Emissions Reduction Targets in the 12th FYP*

The sixth chapter of the Outline deals with green development and constructing a resource efficient and environmentally friendly society. It emphasizes that with ever greater resource and environmental constraints, we must deepen the sense of crisis and establish green and low-carbon development concepts. We should increase efforts around promoting energy conservation and emissions reduction, improving incentive and disincentive mechanisms surrounding environmental development, accelerating resource efficient and environmentally friendly production and consumption patterns, enhancing sustainable development capacity, and improving the overall ecological level of civilization. The 21st chapter of the Outline addresses the strategies to deal with global climate change from three angles: controlling greenhouse gas emissions, enhancing adaptive capacity, and participating in wider international climate cooperation.

Resource and environment related indicators account for a large proportion of indicators used by the 12th FYP to determine economic and social development, including land use, water use, energy conservations, low-carbon development, emissions of major pollutants, as well as forestation and forest management (see Table 4.1). Three binding indicators highlight energy conservation and low-carbon development: reducing energy consumption per unit of GDP by 16%, reducing carbon dioxide emissions per unit of GDP by 17%, and increasing the non-fossil fuel energy share of primary energy consumption from the 2010 level of 8.3% up to 11.4% by 2015 (an average annual growth 3.1%).

The 12th FYP targets demonstrate a clear and comprehensive consideration of the promise made to the international community and of domestic strategic requirements for promoting economic restructuring and transitioning between growth patterns. Compared with the single energy-saving target found in the 11th FYP, these objectives are much more systematic and comprehensive and more directly and closely linked to climate change.

As a major GHG emitter, China is facing growing pressure from the international community. Before the Copenhagen conference in 2009, the Chinese government announced its Nationally Appropriate Mitigation Actions (NAMAs) targets: to reduce carbon dioxide emissions per unit of GDP by 40%–45% of 2005 levels by 2020; to increase the share of non-fossil fuels in primary energy generation to 15%; and to increase forested area to 40 million hectares and forest stock to 1.3 billion cubic meters above 2005 levels. Premier Wen Jiabao, on behalf of the Chinese government, stressed that "we must be true in word and resolute in deed." Irrespective of the negotiations' outcome, China will undoubtedly put forth its greatest effort to achieve or excel above the targets it has set for itself.

The 12th FYP is crucial for accelerating the shift in China's mode of economic development and in order to restructure its economy. Policies and measures aimed at energy conservation and emissions reduction are key to adjusting the economic structure, increasing domestic demand, and promoting sustainable

TABLE 4.1 *Economic and social development indicators related to resources and environment in the 12th FYP*

Indicators		2010	2015	Annual average growth (%)	Type
Area of cultivated land (100 million mu*)		18.18	18.18	0	binding
Reduction of water consumption per unit of added industrial value (%)				30	binding
Efficient utilization coefficient of agricultural irrigation water		0.5	0.53	0.03	anticipatory
Share of non-fossil fuels in primary energy consumption (%)		8.3	11.4	3.1	binding
Reduction of energy consumption per unit of GDP (%)				16	binding
Reduction of carbon dioxide emissions per unit of GDP (%)				17	binding
Reduction in emission of primary pollutants (%)	COD			8	binding
	SO_2			8	
	ammonia			10	
	NO_X			10	
Forest	Forest coverage (%)	20.36	21.66	1.3	binding
	Stocking volume of forest (100 million cubic meters)	137	143	6	

* mu, a unit of area (= 0.0667 hectares)
Source: National economic and social development of 12th Five-Year Plan, March 2011.

development. With that in mind, promotion of green, low-carbon development has become universally accepted.

Despite a consensus on the direction of development, different research institutions and scholars have given different figures for energy conservation and low-carbon development targets for the 12th FYP. Some advocated maintaining a high energy conservation target of reducing by 20%, while others preferred a more moderate 15%. The final target of 16% strikes a compromise between various options, and the carbon intensity target of 17% is one percentage point greater to reflect increased utilization of non-fossil fuel energy.

B *Challenges to Achieving the 12th FYP Targets*

The energy conservation and low-carbon targets proposed in the 12th FYP are by no means easy to fulfill. Based on the challenges faced by efforts directed at the 11th FYP targets, there will be many tough challenges to overcome.

First, China is still in the process of rapid economic growth and accelerating urbanization. Although the 12th FYP set the goal for economic growth at only 7%, economic growth goals at the local level remain markedly higher. Over the next five years, the rate of industrial energy consumption is likely to grow more slowly, but the demand for transportation, buildings, and residential energy use will continue to grow rapidly. A decrease in one factor alongside an increase in three is likely to result in intense growth of total energy consumption.

Second, it is difficult to fundamentally change China's coal dominated energy mix. According to expert estimates, for the share of coal in the energy mix to drop by one percent, about 40–60 million tce other alternative energy will be needed. Although wind, solar, and other renewable energy generation has been growing rapidly in recent years, it remains difficult to increase the share of non-fossil fuels in primary energy generation as total energy consumption also increases quickly.

Third, during the 11th FYP period, phasing out antiquated production capacity played a crucial role in meeting energy conservation targets. As the marginal abatement cost increases, the so-called "low-hanging fruit" (low-cost emissions reduction opportunities) will gradually disappear. Hence, although the 16% reduction target in the 12th FYP is lower than that of the 11th FYP, it might nevertheless be more difficult to achieve.

Finally, achieving the binding energy conservation target of the 11th FYP relied in great part on the government's administrative power. Energy conservation was still lacking effective policies and mechanisms. Some problems, such as a "one size fits all" mentality used to distribute targets to local governments and enterprises were already present. Under the high pressure applied through administrative penalties for noncompliance, some local governments and enterprises that found it particularly difficult to meet targets were forced to employ irrational strategies to reduce energy consumption, such as simply cutting off the electricity, causing highly adverse effects on normal production and residential daily life. During the 12th FYP period, the government must improve target distribution, implement relevant policies, and adopt a more comprehensive policy portfolio of fiscal and financial instruments to better mobilize the entire society.

II Key Polices and Measures for Promoting Energy Conservation and Low-Carbon Development Targets as well as Disaggregation in the 12th FYP

The Chinese government has attached great importance to energy conservation and low-carbon development. In order to achieve the targets set in 12th FYP, a series of policies and measures have been adopted and the targets have been distributed among regions and industries.

A *A Comprehensive Program of Energy Conservation and Emissions Reduction for the 12th FYP*

In August 2011, the state council unveiled a comprehensive program of energy conservation and emissions reduction for the 12th FYP. The program emphasized that all levels of government and all departments should raise their awareness with regard to the urgency of energy conservation, in order to increase global awareness of the crisis, to raise global sentiment of responsibility, to promote a green and low-carbon development concept, and to strictly implement commitments that are made. It is crucial to build a system that is led by the government, driven by market force resource allocation, and with a wide participation by enterprises and full society stakeholders. It is similarly important to enhance leadership, institutional supervision and inspection, as well as to enforce stricter evaluation and accountability.

The program further defined the base year and the energy conservation targets for the 12th FYP, which are to reduce energy consumption to 0.869 tce per 10 thousand RMB of GDP (prices in 2005) by 2015, about 16% below the 1.034 tce in 2010 and 32% below the 1.276 tce in 2005. If the targets are met, 670 million tce can be saved during the 12th FYP period.

In order to achieve these targets, the program put forth 50 specific policies and measures spread across 12 general categories, for example, to strengthen the responsibilities for energy conservation and emissions reduction targets, to adjust and optimize industrial structures, to implement key energy conservation projects, to strengthen energy conservation auditing and management systems, to accelerate low-carbon technology R&D and deployment, to improve economic instruments, to tighten supervision and inspection, to build market mechanisms, etc. Compared with the policies of the 11th FYP, the 12th FYP policies and measures reach farther and deeper in the following ways.

1 Controlling Total Energy Consumption

The 12th FYP maintains the 11th FYP intensity target, but but seeks to control total energy consumption as well to prevent local governments from pursuing higher GDP growth just to increase their denominator. The program suggests a cap of total energy consumption, a breakdown of the objectives among local governments, a clarification of local government management responsibility, and enhancement of supervision and inspection. Energy Saving Assessments for fixed assets investment projects will be an important measure for controlling primary energy consumption on a regional level.

2 Optimize the Industrial and Energy Structures

The driving idea of the 12th FYP is to facilitate and accelerate structural adjustment. The program requires a continuation of the efforts to curb the excessive growth of energy and emissions intensive industries, make a working plan for phasing out obsolete production capacity, distribute tasks to local governments annually, and strictly implement the "Guideline Catalog for Industrial Structure Adjustment" to promote the transformation and upgrade of traditional industries. The share of value added services and strategic emerging industries in GDP will reach 47% and 8% respectively in 2015. The share of non-fossil fuels in total primary energy consumption will reach 11.4%.

3 Implementing Key Energy Conservation Projects

During the 11th FYP period, investment in the Ten Key Energy Conservation Projects produced great environmental benefits, saving a total of 340 million tce, exceeding the expected 240 million tce. During the 12th FYP period, Key Energy Conservation Projects will allow savings of 300 million tce. Compared with 2010 levels, the average operating efficiency of industrial boilers and furnaces will be improved by 5% and 2% respectively, and motor system efficiency will increase by 2–3% by 2015. New waste heat and pressure generation capacity will increase by 20 GW. In the buildings sector, central heating system in more than 400 million square meters of existing residential buildings in northern areas will be retrofitted with thermal meters and for greater energy savings, 50 million square meters of existing residential buildings in hot summer and cold winter areas will be also be retrofitted, as well as 60 million square meters of public buildings. In addition, the market share of energy-efficient products will go up significantly.

4 Promoting Energy Conservation and Emissions Reduction in All Areas

The program stipulates that industrial companies will be assessed by energy standards of activity. Energy conservation audits and management programs for the top 1000 largest energy users in the 11th FYP period will be expanded to 10,000 enterprises during the 12th FYP. All companies with annual energy consumption above 10 thousand tce will be considered large energy users for the purpose of energy conservation audits and management, saving 250 million tce. In the building sector, the green building action plan and the "energy conservation and warm house" program will be developed and implemented, while initiatives will also be pushed so that 10 thousand enterprises in the transport sector can save energy.

5 Accelerating Low-Carbon Technology Development and Deployment

The program emphasizes the role of low-carbon technologies for achieving energy conservation and emissions reductions. It urges increased investment in basic scientific research of common issues and frontier technologies, as well as demonstration of key technologies and commercialization of equipment. It calls for an energy conservation and emissions reduction technology policy. National catalogs of key energy saving technologies and environmental protection technology will be issued to spur deployment.

6 Improving Economic Policies and Encouraging Institutional Innovation

The Program emphasizes using not only administrative powers but also economic instruments for incentives for energy conservation and emissions reductions in the 12th FYP period. These include rational pricing of resource products; aggressive pricing of residential electricity and water; implementation of differentiated electricity prices and punitive fines; resource tax reform; and other incentives that promote a more efficient and comprehensive utilization of resources. The financial system should develop and adopt more financial instruments to support low-carbon projects, such as a favorable loan and green rating system. Other economic policies and innovative mechanisms mentioned in the program include a certification and labeling system for energy efficient and environmentally friendly products, energy generation dispatch and demand side management, energy service companies (ESCo), emissions permits as part of an emission trading scheme (ETS), and a carbon emissions trading pilot.

7 Strengthening Supervision, Inspection, and Related Capacity

To avoid the imperfect energy statistics, monitoring, and assessment that occurred during the 11th FYP, the program stresses the necessity of pushing through legislation and regulations dealing with energy conservation and emissions reduction, implementing strict energy conservation and environmental impact assessments, enhancing management, supervision, and enforcement, as well as expanding comprehensive systems that cover the three governmental levels: provinces, cities, and counties.

8 Promoting Participation throughout Society

The program encourages regions to explore low-carbon development methods that are local circumstance specific by employing pilot programs. Local governments should lead by mobilizing all kinds of resources and all stakeholders, including communities and households, youth, enterprises, schools, military camps, rural areas, the media, etc. Energy conservation and emissions reduction should be integrated into social value education and public awareness campaigns.

B *Regional Distribution of Energy Conservation Targets*

The comprehensive program of energy conservation and emissions reduction for the 12th FYP includes regional targets. The 31 provinces were divided into five categories according to the socioeconomic development levels and specific circumstances. For example, Group 1 has the highest target at 18%, while the lowest target is 10%.

TABLE 4.2 *The regional energy conservation targets for the 12th FYP*

Group	Regions	Reduction of energy consumption per unit of GDP
1	Tianjin, Shanghai, Jiangsu, Zhejiang and Guangdong	18%
2	Beijing, Hebei, Liaoning and Shandong	17%
3	Shanxi, Jilin, Heilongjiang, Anhui, Fujian, Jiangxi, Henan, Hubei, Hunan, Chongqing, Sichuan and Shanxi	16%
4	Inner Mongolia, Guangxi, Guizhou, Yunnan, Gansu and Ningxia	15%
5	Hainan, Xizang, Qinghai and Xinjiang	10%

Source: State Council, August 2011.

Compared with the 11th FYP, where most provinces were given an average target of 20%, the distribution of targets has greatly improved. After multiple rounds of bargaining between central and local governments, the outcome reflects past efforts in the 11th FYP period, regional disparities, and industrial patterns. That explains why the fifth group received a target significantly lower than the other four groups. Considering their relatively low development level and limited mitigation capacity, the lower target left some space for further economic development.

C *Regional Distribution of Non-Fossil Fuel Development Targets*

The distribution of non-fossil fuel development targets for the 12th FYP has not yet been officially issued, but the underlying goals have been disclosed. The expected development goals predict that total primary energy consumption will be 3.63 billion tce in 2015, breaking down into 3.82 billion tons of raw coal (2.610 billion tce); 500 million tons of oil (710 million tce); and 230 billion cubic meters of natural gas (310 million tce). The share of coal in the energy mix will drop from 70.9% in 2010 to 63.6%, a 7.3% reduction. In 2015 non-fossil fuel use will reach 470 million tce, broken down into 280 million tce of hydropower, 90 million tce of nuclear power, and 100 million tce of other renewable energy sources including wind, solar, and biomass. The share of non-fossil fuels in total primary energy consumption will be 11.5%, 0.1% higher than the target in the 12th FYP.

To guide development and utilization of non-fossil fuel development in different regions, special guides with precise numbers have been issued to each province for wind power and nuclear power development along with the 12th FYP distributed energy conservation targets.

1 Wind Power

During the 11th FYP period, wind power installed capacity significantly exceeded national plans. Access to the grid has been a serious barrier to further development of wind power in some regions. In August 2011, the National Energy Administration made a plan for wind power projects construction management. The total new installed wind power capacity during the 12th FYP will be restricted to 25.83 GW. The first batch of new wind power projects which will produce 12.83 GW has been announced and distributed among 25 provinces. For example, provinces with abundant wind power resources received relatively higher quotas, such as 2.1 GW for Heilongjiang, 5.63 GW for Inner Mongolia, 3.02 GW for Hebei, 2.08 GW for Jilin, and 2.3 GW for Xinjiang.

These new wind power project quotas are generally lower than what provinces were expecting. Due to a shrinking market, the wind power industry has shifted from expansion to integration. Decreasing profits threaten the survival of many small wind power enterprises. Small and medium-sized wind power equipment manufacturers have to compete with each other to sell of stock and get new orders.

According to previous wind power management regulations, investment in a wind farm above 50 MW needs to be ratified by the National Development and Reform Commission (NDRC). Wind farms under 50 MW can be ratified by local governments and reported to the NDRC. The flexible procedure allowing local government approval stimulated rapid development of wind power in recent years but also led to blind investment and waste. The plan and quotas for new wind power projects in the 12th FYP reduced this flexibility in order to focus on development of wind power efficiency and quality and to guide the wind power industry through healthy development.

2 Nuclear Power

The Fukushima nuclear accident in March 2011 induced great concern throughout the international community about nuclear security. It also demonstrated to the world that nuclear danger does not respect national boundaries, and that human beings can be easily engulfed by a nuclear disaster. After the accident, China worked with the international community to strengthen monitoring, analysis, and genuine disclosure of information to the public. At the same time China initiated a comprehensive security inspection on all nuclear facilities, enhanced management of nuclear plants in operation, reevaluated nuclear power plants under construction, drew up nuclear security plans, adjusted and improved medium and long term nuclear power development planning, and tightened examination of new applications for nuclear power projects. All these policies and measures played a very important role in improving government transparency and enhancing public confidence in nuclear power. However, since nuclear power is an important option for dealing with global energy shortages and climate change, China insists on developing nuclear power. Under the circumstances, greater security measures will be adopted and nuclear development targets will be much more prudent. The latest data show that China is current constructing 28 units of nuclear power with a capacity of about 40 GW in total. These plants will begin operation in the next five years. The 40 GW target for 2015 will therefore be fulfilled without building any more inland new nuclear power plants in the near future. The share of nuclear power in total primary energy consumption will be decreased to less than 3%.

D *Distribution of Targets for Phasing Out Obsolete Production Methods*

Phasing out obsolete production methods is an important measure for industrial structural adjustment, energy conservation, and emissions reduction, as well as for industrial competitiveness and regional economic development. The Ministry of Industry and Information Technology has issued tasks in 2011 and distributed the targets among industries and regions in accordance with previously established targets, requirements for industry upgrades, and specific circumstances in different industries and regions.

The targets for phasing out obsolete production deal with 18 energy intensive industries that make up 15%–25% of total production, for example, 26.53 Mt and 26.27 Mt respectively for iron and steel smelting, 1,813 Mt for nonferrous smelting (aluminum, copper, lead, zinc), 133.55 Mt for cement, 7445 tons for the paper industry, etc.

Different regions committed to phase out obsolete production in different ways. For example, the provinces Hebei, Henan, Shandong, and Shanxi will focus on the iron, steel, coke, flat glass, and paper industries. Hebei, Shanxi, Liaoning, and Zhejiang will focus on the cement industry. Obsolete printing and dying industries are mainly located in Hubei, Shandong, and Zhejiang.

III Regional Initiatives Promoting Energy Conservation and Low-Carbon Development

The central government distributed the 12th FYP energy conservation and low-carbon targets to industries and regions using a top-down approach, while regions proposed their own targets and working plans for the next five years using a bottom-up approach.

Control of total primary energy consumption serves as a good example for demonstrating the interaction between the central government and provincial governments around different targets and working plans.

In May 2010, nine departments including the Ministry of Environmental Protection, the National Development and Reform Commission, and the National Energy Administration issued a notice meant to guide and improve regional air quality using comprehensive prevention and control methods. The first step was to control total coal consumption in pilot zones, which include the three regions of Beijing, Tianjin, and Hebei, the Yangtze and Pearl River Deltas, and six urban agglomerations in the central part of Liaoning, the Shangdong Peninsula, Wuhan, Chang-Zhu-Tian, Chengdu-Chongqing, and the west side of Isthmus. These regions and urban agglomerations are all located in areas with rapid economic growth. In order to achieve the target, each region determined its own targets and related policies and measures. For example:

Beijing surpassed the 11th FYP energy conservation target, ranking no.1 in its group. Because Beijing cannot relocate the same energy intensive enterprises or re-host the Olympic Games during the 12th FYP, the 17% target set for this group 2 city presents many challenges. Wishing to multitask and work on energy conservation, carbon reduction, and air pollution simultaneously, Beijing put coal consumption control as its top method for controlling energy consumption. Beijing is the first city to set a total coal consumption target. According to the 12th FYP energy development plan and the "Beijing Clean Air Action Plan" (2011–2015) announced in August 2011, Beijing set average annual economic growth for the 12th FYP at 8%. The total energy demand will be about 85–95 million tce in 2015, 30% above 2010 levels, and the annual average growth rate is about 5.34%. Beijing plans restrict total coal consumption to no more than 20 million tons in 2015, 7 million tons below 2010 levels. Meanwhile, Beijing plans to increase the share of natural gas in total energy consumption to 20%.

Shanghai, which belongs to Group 1, has to reduce energy consumption per unit of GDP by 18% during the 12th FYP period. As opposed to Beijing's plan to control total coal consumption, Shanghai prefers to control total energy consumption. Shanghai plans to restrict total energy consumption to no more than 140 million tce in 2015.

Tianjin, which also belongs to Group 1, is likely to adopt a plan to control total coal consumption. Lacking natural gas, it is difficult for Tianjin to follow Beijing's absolute reduction of coal consumption. Although Tianjin set a target to restrict coal consumption at a level that is 49 million tons greater than 2010 levels, direct combustion of coal will be decreased using coal gasification and other measures.

Jiangsu Province, *Guangdong* Province and *Zhejiang* Province, all of which belong to Group 1, have set total energy consumption targets at 341, 228, and 320 million tce respectively. Jiangsu Province has placed the emphasis on increasing energy supply, optimizing energy structures, accelerating science and technology development as well as institutional innovations, increasing energy efficiency, and building modern energy systems that are safe, stable, economic, and clean. Jiangsu has proposed four objectives for 2015: (1) to restrict total energy consumption to no more than 341 million tce; (2) to increase the share of non-coal fuels to 30%, natural gas to 10%, nuclear power to 1.4%, and wind power to 1.3%; (3) to reduce energy intensity to 18% below 2010 levels, reaching 0.6 tce per 10 thousand yuan (price in 2005 RMB); (4) to reduce coal consumption for electricity supply to 317 g/kWh, and reduce carbon emissions intensity to 19% below 2010 levels.

Fujian Province, which belongs to Group 3, has to reduce energy consumption per unit of GDP by 16% during the 12th FYP period. In accordance with the 12th FYP energy plan, Fujian set a series targets in 2015: to restrict total energy consumption at no more than 140 million tce, an annual growth rate of 7.6%. Electricity consumption will be held to 227 billion kWh and peak load to 38.4 GW, an annual growth of 11.5% and 11.6% respectively.

Hunan Province, which also belongs to Group 3, set targets to restrict energy consumption to no more than 230 million tce, and decrease energy consumption per unit of industrial value added by 18%. Unlike other regions, Hunan emphasizes key construction projects. Hunan plans to build 77 key construction projects related to energy conservation and carbon reduction, ecological governance, industrial energy efficiency, comprehensive management of the Xiangjiang River valleys, ecological and environmental protection etc. Total investment for these projects reaches 490 billion RMB, 450 billion RMB just in the five upcoming years.

Jilin Province originally set the highest energy conservation targets at 30% but then adjusted to 22% later on during the 11th FYP period. Jilin, which belongs to Group 3, seems to be taking a much more pragmatic attitude to the 12th FYP. It initially set a target to restrict total energy consumption to no more than 135 million tce in 2015.

Yunnan Province, which belongs to Group 4, has committed to reduce energy consumption per unit of GDP by 15% during the 12th FYP. Taking advantage of abundant hydropower resources and other renewable resources, Yunnan set a target to reduce carbon emissions per unit of GDP by 20%, surpassing the national goal of 17%. However, Yunnan has not set a target to restrict total energy consumption.

The *Xinjiang* autonomous region made great strides on energy conservation and carbon reduction during the 11th FYP. Energy consumption per unit of GDP decreased by 8.9%, successfully reversing the upward trend of the 10th FYP period, however failing to achieve the designated target. During the 12th FYP period, Xinjiang belongs to Group 5, with an energy conservations target set just at 10%, but even this will not be easy to achieve. Xinjiang emphasizes strategies that put environmental protection and ecological construction at the top. Yunnan has not set a target to restrict total energy consumption.

IV Outlook and Policy Recommendations for Achieving Energy Conservation and Low Carbon Development in the 12th FYP

The 12th FYP period is crucial for raising the standard of living, expanding reform and liberalization policies, and accelerating the transition of economic development. In order to successfully achieve energy conservation and low-carbon development targets, especially increasing the share of non-fossil fuels in primary energy consumption, it is necessary to control the total primary energy consumption. During the 11th FYP period, some regions reduced energy intensity exclusively by growing the denominator, which does not advance low-carbon development. This explains the shift from energy intensity control to total energy consumption control for the 12th FYP period. However, the transition presents quite a few obstacles and difficulties.

First, the GDP growth rate target does not align well with total energy consumption control targets at the national level. The expected GDP annual growth rate during the 12th FYP period is 7%. If the target is met, a 16% reduction in energy intensity will result in total energy consumption of about 3.8 billion tce in 2015, which appears impossible. If the GDP growth rate were set around 8%, the total energy consumption target would be around 4.1 billion tce. In reality, GDP growth rate reached 9.2% in 2011. In order to set a feasible target for total energy consumption, it is likely that the GDP growth target will not be respected.

Second, national top down targets are not distributed in accordance with regional bottom up targets. According to research estimates, a reasonable total energy demand for 2015 is around 4.0–4.3 billion tce, and the total energy cap was set at 4.1–4.2 billion tons. However, the sum of regional energy consumption caps will perhaps reach 5.0 billion tce. Aside from inconsistent statistics, the mismatch further also reflects that the majority of local governments are eager to increase energy consumption for local economic development. A national distribution of caps is extremely challenging and bargaining and lobbying are sure to happen in the process. It might be one of reasons for the delay in publishing the Energy Development Plan of the 12th FYP, which was finally passed by the State Council on Oct. 24, 2012.

Third, in the process of implementing the 12th FYP energy conservation and low-carbon development targets, different industries may bargain to protect their interests. For example, fossil fuel industries, non-fossil energy industries, coal-burning power generation plants, renewable energy projects, and others, may find themselves in competition. Energy intensive industries such as steel and cement are more sensitive than others. Thus, total energy consumption caps have to take their special demands into account. Enterprises are playing an ever greater role in decision making processes. The relationship between enterprises and the government has gradually shifted from passive obedience to proactive interaction.

Fourth, if total energy caps are distributed among regions and industries, enforcement mechanism are still unclear. Administrative power is effective but inflexible, and suffers from high economic costs. Extreme measures such as cutting electricity supply to reduce energy consumption might reoccur if great pressure is applied. Market-based mechanisms such as emission trading schemes (ETS) help to reduce economic costs and increase flexibility for enterprises and the government. Some conditions may allow a CO_2 ETS to be introduced during the 12th FYP period, but there are many policies which must still be carefully crafted and the project can only be advanced step by step owing to China's imperfect market economy.

Finally, to implement total energy consumption caps requires a series of institutional innovation, such as energy statistics and greenhouse gases inventory, accounting and reporting systems, codes and standards for energy efficient and low-carbon products, labeling and certification systems, evaluation and assessment systems, reward and punishment systems, energy conservation assessment for fixed asset investment project and review systems, etc.

The 12th FYP begins in 2011. 2012 is important for staying on track. According to the latest statistics available, energy intensity only decreased by 2.01% in 2011, much less than the 3.5% annual target. The share of non-fossil fuels in total primary energy consumption failed to increase, and in fact decreased from 8.6% to 8%. Numbers seems better in 2012. The energy consumption per unit of industrial value added decreased by 7.35% from January to May of 2012. A slowing of economic growth may be the lead reason. The current situation indicates that the tasks necessary for achieving the 12th FYP targets over the next three years will be greatly challenging.

Overall, the global trend of green and low-carbon development brings both serious challenges and important opportunities to China. Successfully achieving the 12th FYP targets is not only China's promise to the international community, but also its own choice in order to transition economic mode and restructure sustainable. China should ensure that is has satisfactory results to show its people and the world.

(This article was originally published in Chinese in 2012.)

CHAPTER 5

Progress of Low-Carbon Urban Construction in China

Zhuang Guiyang and Wang Chunli

Abstract

Low-carbon urban construction in China is transitioning from its initial spontaneous and sporadic attempts to national level systematic arrangements. The goal is to form a scientific low-carbon economic development framework. The low-carbon pilot project "Five Provinces and Eight Cities" led by the National Development and Reform Commission, the low-carbon transportation system pilot project led by the Ministry of Transport, as well as integrated demonstrations of energy saving and emission reducing fiscal policy from the National Development and Reform Commission and the Ministry of Finance, all follow a similar pattern of demonstrations through pilot projects, summary of experiences, and gradual promotion. This pattern appears to be Chinese policy makers preferred method. Low-carbon urban construction in China incorporates advanced planning, and accelerated improvements to accounting systems for greenhouse gas emissions statistics.

Keywords

low-carbon city – pilot demonstration – progress – focus

I Practical Exploration of Low-Carbon Urban Construction in China

Low-carbon development is a new concept, which was invented for global climate change adaptation in recent years. At its core it seeks to reduce carbon emissions from development processes to guarantee global temperature

---

* Zhuang Guiyang, Senior Research Fellow at the Institute for Urban and Environmental Studies at the Chinese Academy of Social Sciences. Research field is low-carbon economy and climate change policy. Wang Chunli, Senior Associate of Ernst & Young (China) Advisory Limited Beijing Branch office.

 DOI 10.1163/9789004274648_006

rise no more than two degrees above per-industrial temperatures. Since cities accounts for over 70% of total emissions (local emissions and indirect emissions caused by consumption), cities concentrate people and economic activity, low-carbon urban development has become an important goal for climate change adaptation in every country. As developed countries and international metropolises are actively transitioning to low-carbon economies, cities are also participating actively. Since 2008, over 100 cities such as Baoding, Shanghai, Guiyang, Hangzhou, Dezhou, Wuxi, Jilin, Zhuhai, Nanchang, and Xiamen, have proposed incorporating low-carbon urban construction, and so all have become national low-carbon pilot cities. Although each city is free to interpret "low-carbon," there is no doubt that cities are thinking and exploring new urban development patterns that reflect the city's economic and competitive position.

A *Local Styles*

As the largest and city in China with the greatest level of imports, it is difficult for Shanghai to make a comprehensive low-carbon transition across all levels. However, the urban low-carbon practices in Shanghai reveal its focus on operational effectiveness. Although none of its plans have been fully implemented, Shanghai has not shied away from partial implementation. The 2010 Shanghai World Expo site became a "low-carbon experimental field" for Shanghai. With an aim to reduce carbon emission at the source, Shanghai took the World Expo site construction as an opportunity to implement low-carbon concepts throughout the entire process, such as location selection, planning, design, construction, operation, etc. In addition, the World Expo Organizing Committee actively implemented carbon offset measures where possible to offset extra carbon emissions generated by the World Expo. The Shanghai World Expo took many of the world's most advanced low-carbon technologies to implement low-carbon transportation, low-carbon construction, low-carbon energy generation, low-carbon integrated city, etc. It also spread the concepts of low-carbon production and low-carbon life. The World Expo site provides systematic solutions and an experimental playing field for future low-carbon urban development projects through comprehensive application of low-carbon technology. The Shanghai World Expo has a strategic effect, technical effect, and conceptual effect on low-carbon development in Shanghai's future.[1]

1 Yu Hongyuan, Tang Wei 于宏源、汤伟, "Ditan shibo yu chengshi fazhan 低碳世博与城市发展 [Low-carbon Expo and Urban Development]," see Wang Weiguang, Zheng Guoguang 王伟光、郑国光, "(Yingdui qihou bianhua baogao 2010 应对气候变化报告 2010 [Climate Change Initiative Report (2010)]," Social Science Academic Press, 2010.

B *Featured Industry Development*

Compared with Shanghai, some cities have obvious advantages that will allow them to use low-carbon industries in low-carbon development and raise competitiveness. Baoding, for example, followed Silicon Valley in California's development model and proposed to construct a "China Power Valley." In recent years, six industries have emerged: photoelectric, wind power, energy conservation, power storage, power transmission, and power automation. Baoding promoted the innovation and application of new energy production technologies and used it as a foundation for a low-carbon city. Dezhou City has abundant solar energy resources, so the city government has decide to promote solar energy. In 2005, Dezhou City implemented the "China Solar City" strategy as well as several city-scale photoelectric lighting demonstration projects such as "million roofs," "thousand village bathrooms," and "5555" to promote the wide use of solar energy. Solar light and heat technology is well developed and popularly accepted, leading to broad use in Dezhou. The urban household penetration rate of solar water heaters in Dezhou is 75.4%, and the rural household solar water heater penetration rate is around 15.0%. The solar industry in Dezhou has developed rapidly and it is occupying a leading global role. It has built a relatively complete industrial chain that includes vacuum tube solar water heaters, solar models, photovoltaic systems, solar lighting systems, solar traffic lights, energy-saving glass, and integrated building solar technology.[2]

C *Systematic Planning*

Many cities would like to develop a low-carbon economy, but some cities are weighed down with projects from the past, trapped with high pollution, high energy consumption, and with low value-added industries. Some are located far from development centers and opportunities. A number of Chinese cities will have to seek a new strategy to break past these struggles. The old industrial city of Jilin in northeastern China was designated by the National Development and Reform Commission to be the first pilot city for a low-carbon economic case study. It provides experiences and practical support for national policy making. In March 2010, the results of the "Jilin City Low-Carbon Development Roadmap"[3] were officially released. Jilin City identified

2 Environmental Policy and Environmental Planning Institute Renmin University 中国人民大学环境政策与环境规划研究所, "Dezhou shi 'shi'er wu' ditan fazhan guihua yanjiu baogao 德州市'十二五'低碳发展规划研究报告 [Dezhou City 12th FYP Low-carbon Development Planning Study Report]," 2010.12.

3 Low-Carbon Development Roadmap for Jilin City, Chatham House, Chinese Academy of Social Sciences, Energy Research Institute, Jilin University, E3G, 2010.

key industries from among its existing industries which are expected to support the industrial transformation, with the petrochemical industry clearly positioned as a central pillar. Along with economic development, Jilin City has focused on improving industrial structures and reducing energy intensity to reduce carbon output. Hangzhou City faces a different situation. In July 2008, the Hangzhou CCP Municipal Plenary Session of the 10th Four Year Plan, proposed to make carbon emissions reduction a higher priority, develop a low-carbon city indicator system, an evaluation system, and a special action plan, build a low-carbon technology museum, build a low-carbon economic demonstration area, actively adjust the energy structure, develop the low-carbon economy, promote a low-carbon lifestyle and build toward a low-carbon city. In 2009, Hangzhou drafted 50 "Low-Carbon New Policies" which clearly stated Hangzhou's willingness to be the first city in the country to build a low-carbon city with a low-carbon economy, low-carbon buildings, low-carbon transportation system, a low-carbon lifestyle, a low-carbon environment, and a low-carbon society.

II Pilots and Demonstrations that Promote Low-Carbon Urban Construction in China

China has adopted a strategy of summarizing the experiences of pilot demonstrations and using them to promote further projects. Low-carbon economic theory and practice has been expanding in China. Although many cities have proposed ideas for low-carbon construction, they are generally spontaneous, sporadic, and tentative, lacking a unified system. It is important to start at easy access points, and then progress to small projects, before launching full experimental initiatives. As the low-carbon concept becomes gradually more widely accepted, some cities are pushing for low-carbon development, some are only proposing the ideas, while some strive for the radical concept of a "Zero Carbon City," which ignores technical and social reality. Certain cities are adopting concepts not directly tied to low-carbon development in order to arrive at low-carbon urban planning, such as a circular economy or a green economy, but in doing so they ignore one of the most critical indicators- carbon emissions. In order to unify knowledge and regulate domestic low-carbon practices, China has carried out low-carbon urban construction pilot demonstrations testing various subjects and concepts.

A *External Force Oriented Development*

Low-carbon economy was first introduced to China as an academic concept when international institutions funded academic groups to research "low-carbon city programs." International cooperative research programs became the early pilot programs and demonstrations of low-carbon city research. In the autumn of 2007, the Rockefeller Brothers Fund supported a Climate Group to do research on a low-carbon economy roadmap for the Pearl River Delta region, including Guangdong and Hong Kong, which directly promoted low-carbon explorations in these two regions. WWF's "Low-Carbon City Development Program" chose Baoding City as their pilot city in the beginning of 2008 because of its attempt with new energy industries, thereby exposing local government leaders to low-carbon concepts earlier than in other cities, which helped to promote low-carbon development in other urban operations. The UK Strategic Program Fund (SPF) promoted low-carbon development in several provinces and regions in China, such as in Jilin City, Nanchang City, Chongqing City and Guangdong Province. With the support of the Energy Foundation, Tsinghua University and other institutions carried out a low-carbon strategy preliminary study for Shandong and Suzhou. Local research institutes in Guangdong, Hubei, Chongqing, Nanchang, Baoding, and other cities continued developing low-carbon pilot programs based on preexisting research. The Switzerland-China Low-Carbon City Program launched in June 2010, and Yinchuan, Beijing Dongcheng District, Dezhou City, and Meishan City were chosen as pilot cities and areas. The program focused on urban management, low-carbon economy, low-carbon transportation, green construction and other fields. Low-carbon development action plans were created to further low-carbon development in these cities.[4]

B *Development Oriented at Overall Promotion*

The National Development and Reform Commission selected five provinces (Guangdong, Liaoning, Hubei, Shaanxi, Yunnan) and eight cities (Tianjin, Chongqing, Shenzhen, Xiamen, Hangzhou, Nanchang, Guiyang, Baoding) in August, 2010 to serve as pilot provinces and cities in order to unify knowledge and further steps in light of the need for local development of low-carbon economies. The selection of these pilot provinces and cities considers preexisting infrastructure for low-carbon development as well as geographical representation model diversity. These pilot provinces and cities have clear regional characteristics, are at different development stages, are composed of different industrial structures, and differ in their resources. This is conducive not only so that the national government can accumulate experiences to guide green, low-carbon development in different regions, but also to create vast spaces for initiative and creativity.

4 Climate Group 气候组织, "Zhongguo de qingjie geming III: chengshi 中国的清洁革命:城市 [China's Clean Revolution III: City]," 2010.

As the pioneers and explorers of green, low-carbon development, these pilot provinces and cities will play a leading and exemplary role in fields such as GHG emissions control. The regions set carbon emission intensity targets to decrease the overall requirements of the central government and also to reflect the advances of pilot areas as exemplary cases. The pilot provinces and cities are instructed to explore the following five aspects: preparation for low-carbon development planning, formulation of policies to support green, low-carbon development, acceleration and establishment of low-carbon industry systems, establishment of GHG emission statistical and management systems, and active promotion of a green, low-carbon life-style and consumption pattern.

C *Development Oriented toward Industry Promotion*

Due to the rigidity of urban planning and construction, once certain infrastructure is built, it is difficult to change in what is known as a "lock-in effect." Urban infrastructure master plans must be drawn up beforehand to ensure the low-carbon design of urban infrastructure. Carbon emissions from the transportation sector are a focal point of developed countries and they represent a potential source for carbon emission growth in China. Experiences around the globe have proved that policy and institutional innovation serve as important driving forces for promoting carbon emission reductions in the transportation sector, key to low-carbon development. To ease traffic congestion, improve urban air quality, ensure the safety of transportation energy use, and promote the sustainable development of transportation, China has established many policies, thereby reducing the transportation sector's carbon emissions. However, China still lacks a systematic low-carbon transportation strategy. Along with low-carbon provinces and cities pilot programs created by the National Development and Reform Commission, in February 2011, the Ministry of Transport chose 10 cities (Tianjin, Chongqing, Shenzhen, Xiamen, Hangzhou, Nanchang, Guiyang, Baoding, Wuhan, and Wuxi) to carry out low-carbon transportation system construction and experimental work. The Ministry of Transport has established a "Guide on Construction of Low-Carbon Transport Systems" and a "Pilot Program Work Plan for Construction of Low-Carbon Transport Systems" to provide systematic guidance and specific strategies for building low-carbon transport systems. This pilot program has six main objectives: to construct low-carbon transportation infrastructure, to promote the application of low-carbon transportation equipment, to optimize transportation system organization and operation, to build intelligent transportation projects, to improve public information services, and to build carbon emissions management systems for the transport sector.

D *Policy Promotion Oriented*

Saving energy and reducing GHG emissions go hand in hand and are important for the transition to a low-carbon economy. In accordance with the "Twelfth Five Year Plan for National Economic and Social Development," the Ministry of Finance and the National Development and Reform Commission decided to carry out comprehensive experiments of financial and fiscal policies to save energy and reduce GHG emissions, as well as adjust the economic structure and prepare of transitioning economic growth. New breakthroughs are anticipated from the integration of financial and fiscal policy and from large financial investments. In June 2011, the Ministry of Finance and the National Development and Reform Commission chose eight cities (Beijing, Shenzhen, Chongqing, Hangzhou, Changsha, Guiyang, Jilin, Xinyu) as the first batch of case studies that aim to establish a green, circular, and low-carbon development concepts. Concepts include pushing energy conservation methods that can be adopted throughout society, using the government as the leader, enterprises as primary actors, and the market as regulator. The projects also seeks to sharply increase energy efficiency in industry, architecture, transportation and other fields, as well as disperse low-carbon technology and promote the large-scale application of renewable energy; reduce the emissions of main pollutants; accelerate development in the service sector; improve the market mechanism for energy performance contracting, and; push pilot cities to the front line of energy conservation and emissions reduction work. The pilot demonstration of financial and fiscal policies has six specific goals: adjust industrial structure in accordance with low-carbon industry practices, develop a clean urban transportation system, promote architectural energy saving, such as green architecture, develop an intensive service industry, improve urban environmental quality through reduction of lead pollutants, and optimize the urban energy structure through the large scale application of renewable energy.

III Focus on Promoting Low-Carbon Urban Construction in China

Both awareness of low-carbon urban development in China and action toward achieving it are developing alongside ideas and actions for climate change adaptation and low-carbon economic development. Low-carbon urban development not only meets the needs of climate change adaptation, but is also consistent with sustainable development advocated by the Chinese government. Ever greater low-carbon strategies, regulations, and policies are introduced by the Chinese government, such as on November 26, 2009, when the State Council proposed to reduce carbon dioxide emission per unit of GDP in 2020 by 40%-45% of 2005 levels, and set that goal as a binding target in national economic and social development medium- and long-term plans. At the same time it developed the corresponding national statistics, monitoring, assessment methods. Low-carbon development has gradually shifted from optional policy to become the default compulsory strategy for every city.

The most important component of carrying out pilot low-carbon demonstrations is to greatly improve local understanding. Each development zone has achieved a more in-depth understanding of relevant concepts and ideas, established corresponding management organizations, and proposed low-carbon development targets, major initiatives, and key tasks. As low-carbon development planning is implemented in pilot provinces and cities, institutional mechanisms that promote low-carbon development will support the boom in low-carbon industrial urban construction.

A *Low-Carbon Urban Development Planning*

Low-carbon urban development planning is an institutional arrangement of urban space and development timing, based on low-carbon concepts and technology in combination with urban spatial planning and urban development planning made under specific economic and social development conditions. Preparation and planning for urban low-carbon development ensures the process is guided properly using a comprehensive view of the development project. Preparation includes identifying low-carbon development tasks, proposing specific measurements, and exploring urban low-carbon development options using clear urban low-carbon development goals.

The current urban planning system in China consists of three principle parts: urban spatial planning, economic and social development planning, and urban and rural land use planning. The common feature shared by all three is that they map out ways of constructing and developing the city, although they do so from different angles. All three seek multiple targets simultaneously, or otherwise work outside of a target framework. Low-carbon urban development plans build construction and development plans on the basis of low-carbon target by contrast. It is oriented toward a single target, developing the city's economy and society in a low-carbon fashion. As the configuration of urban space and land use indicators have an important influence on urban low-carbon development, such planning should use low-carbon economic and social development plans as a guide for making urban spatial requirements and urban and rural land use plans.[5]

---

5 Zhuang Guiyang, Li Hongyu, and Zhu Shouxian 庄贵阳、李红玉、朱守先, "Ditan chengshi fazhan guihua de gongneng dingwei yu neirong jiexi 低碳城市发展规划的功

Borrowing ideas from research and practice of the development of low-carbon cities at home and abroad, especially in combination with the characteristics of urban development in China, it is recommended that when a city carries out low-carbon urban construction, it should be carried out according to the following steps, or at least that these steps be used to guide the conceptualization, planning, and implementation of low-carbon development projects.[6]

Step 1: Develop a Better Understanding of the City's Current Carbon Emissions

A low-carbon city means the urban carbon emissions or emissions strength ratio is lower than it was previously or lower than that of other cities. Regardless, a clear understanding of the city's current carbon emissions is required as a base. A better understanding of the current carbon emissions, major sources of emissions, and emissions reduction potential is also crucial to guiding low-carbon planning and low-carbon proposals. If a city succeeds in clearly understanding the nature of its carbon emissions over the past several years, it can more clearly analyze the relationship and the development trends between carbon emissions and socioeconomic development. A greenhouse gas inventory is the principle method used for long term sequencing research on urban greenhouse gas emissions.

Step 2: Research Future Medium and Long Term Carbon Emissions Scenarios

After developing a firm grasp of the status of the city's carbon emissions, the city should proceed to analyze various medium and long term carbon emissions scenario using economic and social development trends and targets in order to develop carbon reduction targets and construct low-carbon development plans. Scenario analysis is not a forecast, but rather a method of analysis of the problem and its inherent causes and effects. It is a tool for policy evaluation and strategy planning. The city's energy system and the carbon emissions caused by energy consumption forms the core of scenario analysis. With that information the city can look for the ways to control and reduce the growth rate of urban greenhouse gas emissions, the central focus of scenario analysis.

能定位与内容解析 [Functional Positioning and Concept Analysis of Low-carbon Urban Development Planning]," Chengshi fazhan yanjiu (2011): 8.

6 Lei Hongpeng, Zhuang Guiyang, and Zhang Chu 雷红鹏、庄贵阳、张楚, "Bamai zhongguo ditan chengshi fazhan: celue yu fangfa 把脉中国低碳城市发展：策略与方法 [Analyze China Low-carbon City Development; Strategy and Mothodology]" Zhongguo huanjingkexue chubanshe, 2011.

Policy derived from scenario results will be based on significant knowledge and can properly lead urban low-carbon and energy development strategies.

Step 3: Set Quantitative Carbon Reduction Targets

Comprehensive quantitative carbon reduction targets are at the core of low-carbon urban development. Carbon reduction targets should be feasible, taking into full account the city's development stage, resource endowments, emissions layout, development orientation, etc., while simultaneously taking the country's overall goals into account. The targets should also show a willingness to advance and promote the image of the city. Chinese cities are keener to reduce carbon intensity, but there are some cities setting per capita emissions targets as well. Taking into account population changes, per capita emissions targets are also total emissions targets. Setting total emissions target is the inevitable direction for future development. The first cities to boldly accept that challenge will also be the first to benefit from its results.

Step 4: Prepare Urban Low-Carbon Development Plans

The city's low-carbon development plan is a general program of all low-carbon actions; a guiding document that breaks down carbon reduction targets by industry, and explains the arrangements and requirements that ensure the low-carbon targets will be met using various resource inputs and construction systems. The urban low-carbon development plan should include the major categories of buildings, transportation, industry, and energy, and it should lay out government intervention, financial policies, public participation, and corporate participation. Preparation of urban low-carbon development plans should consider coordination with existing urban economic and social development plans as well as any other special plans, and ensure the legality and feasibility of the core objectives and measures through reasonable means.

Step 5: Implement Low-Carbon Development Plan

The implementation of low-carbon development plans should be led by an urban macroeconomic management department that can coordinate various professional management departments, fully mobilize enterprises and the public, and advance comprehensive system building, financial support, technological support, and public opinion advocacy. Plan implementation should start with the easy steps and then move to the difficult ones, starting with sequencing potential investments toward emissions reductions, selecting first projects with low investments and large reductions. The government should set an example, reducing carbon emissions of governmental office buildings and public infrastructure as a first step, and promote carbon reductions at enterprises and throughout society using a governmental guiding role. It should leave resource allocation to the market mechanism and carry out research and experiments on carbon finance, carbon trading, carbon logo, and carbon certification.

Step 6: Assess and Monitor the Carbon Reduction Effect

Cities should assess and monitor the effects various actions have on urban carbon reduction by preparing annual greenhouse gas emissions inventories. The evaluation can make use of the evaluation index system for low-carbon cities. Assessment and monitoring of urban carbon reductions is conducive to amendment and improvements to the low-carbon development plan over time, ensuring the ultimate achievement of low-carbon goals. Steps 1 to 4 aim to develop a scientific low-carbon urban development plan that is based on individual conditions and features without excessive action. The first and second steps are relatively difficult, but are also the most important. Many cities do a simple analysis of the current situation and of different emissions development scenarios, set a quantitative target for carbon intensity reduction, and then formulate and implement a plan, which has achieved positive results.

In accordance with the requirements of the pilot project "Five Provinces and Eight Cities" which is being supervised by the National Development and Reform Commission, pilot provinces have prepared low-carbon development plans and made pilot city implementation plans. Although the low-carbon development plan and implementation plan for pilot provinces and cities have not yet been made public, the internal information shows that the pattern used for making low-carbon plans in pilot cities are similar and basically followed the steps as stated above.

B *Urban GHG Inventory*

The greenhouse gas inventory is a measurable, reportable, and verifiable prerequisite for urban emissions reduction and a principle tool for climate change adaptation and low-carbon transition. It is a basic and urgent tool for making a low-carbon development roadmap. European and American cities in developed countries account for the majority of urban GHG inventories, while Asian and African cities lack data. China suffers from a large research gap concerning GHG inventories and carbon emissions monitoring on the city level. The preparation of urban GHG inventories as well as the theory and methodology of GHG emissions models lack research, technical support, data, and capacity, which led to a bottlenecking of low-carbon urban development in China and severely restricted low-carbon urban development related scientific decision making.

In 1992, the United Nations Framework Convention on Climate Change decreed that every country has an obligation to prepare a GHG emissions inventory and submit national information. In 1996, the Intergovernmental Panel on Climate Change (IPCC) issued the "Guide for National Greenhouse Gas Emissions Inventory" to provide methodology and operational guidance for countries preparing a GHG inventory. Using constant practice and improvements, the IPCC issued a new emissions inventory guide in 2006 which is currently the international standard for preparing national level GHG inventories. In accordance with the requirements of the National Development and Reform Commission on the work of "Five Provinces and Eight Cities," pilot cities need to prepare GHG emissions statistics and management systems. However, China has yet to form a normalized and standardized preparation system for a GHG emissions inventory. The current process adopted when preparing inventory data relies on small scale sample surveys and expert judgment, which does not generally meet requirements. Although many studies on national energy use and carbon emissions have been released and published, there is not much research on a city scale. Methodology covering GHG research in various city department is still developing and in continuous need of improvement.

Preparation of a GHG inventory is fundamental for addressing climate change. Using a GHG inventory, the major sources of GHG can be identified, emission norms from various departments can be understood, and future mitigation potential can be better predicted, which allows coordination with national and regional GHG emissions control policies and actions. The National Development and Reform Commission is working on the preparation of the 2005 and 2007 national greenhouse gas emission inventories. the core difficulty for the research is defining macro-level data, collecting and testing emission factors, establishing a GHG inventory database management system, and improving the user interface of the database management system, the list query system of the database, and the design of the inventory preparation database.

In order to accumulate experience at preparing provincial level inventories, the National Development and Reform Commission confirmed Guangdong, Hubei, Liaoning, Yunnan, Zhejiang, Shaanxi, and Tianjin as pilot provinces and cities charged with preparing the 2005 GHG emissions inventory. According to the working plan, pilot regions needs to complete a draft of the inventory before June 2011, and finish the inventory report before the end of 2011.

The National Development and Reform Commission requires pilot provinces and cities to ensure the funding for preparation and promptly start collecting basic information and data for preparing a provincial level GHG inventory. The Department of Climate Change of the National Development and Reform Commission recently issued the "Guidelines for the Preparation of Provincial Level Greenhouse Gas Inventory (Trial Version)," which aims to strengthen the preparation of provincial inventory, making it more scientific, normalized, and feasible. It provides useful guidance for a provincial GHG inventory with scientific methodology, transparent data, consistent format, and comparable results.

The preliminary research in several cities shows that work on energy statistics at city level is still at the beginning stages. The Bureau of Statistics in some cities has an Energy Statistics Branch, but many cities have never prepared an energy balance sheet. Work related to urban carbon emission statistics is not well integrated. The preparation and capacity construction of a carbon emissions inventory at the city level will be a long process.

(This article was originally published in Chinese in 2012.)

CHAPTER 6

Status of the Chinese Trading Scheme for Carbon Credits and Future Prospects

Hongbo Chen, Wei Lin and Zheng Zhou

Abstract

Given the strife ridden progress of international talks on climate change mitigation, uncertainty remains regarding the continuation of the Clean Development Mechanism (CDM) post-2012. A number of countries have nevertheless determined market mechanisms as an important approach against climate change. China's 12th Five-Year Plan on National Economic and Social Development clearly states that China will set up and improve the greenhouse gas emissions statistical accounting system and will gradually establish an emissions trading scheme. It is expected that in the upcoming couple of years, China will develop its own policies on carbon trading and gradually establish a domestic emissions trading market. The construction of China's carbon market will progress from a voluntary to mandatory and regional to national manner. China will first standardize and promote the trading of voluntary emission reductions, and will then encourage pilot provinces and cities as well as some key industries to implement policies on emission allowance trading while facilitating the construction of carbon market operation facilities and infrastructure including climate change legislation and carbon emissions measurement and verification systems. China will also carry out international or bilateral cooperation with foreign carbon trading schemes contingent upon the progress of international negotiations, so as to establish a unified domestic carbon market that is linked to international carbon trading schemes.

Keywords

China's carbon market – Status – Prospect – CDM

---

* Hongbo Chen, associate research fellow at the Chinese Academy of Social Sciences (CASS), doctor. He specializes in analysis of climate change mitigation policies as well as in research on carbon markets and low-carbon cities. Wei Lin, General Manager of Easy Carbon. Zheng Zhou, graduate student at the graduate school of CASS.

 | DOI 10.1163/9789004274648_007

I Status of the Chinese CDM Market and Present Challenges

A *Status and Features of CDM Project Development in China*

As of September 15, 2011, China has successfully registered 1,574 projects at the UN CDM Executive Board (EB), accounting for approximately 45.7% of all registered projects among all host countries. These Chinese projects are expected to produce annual savings of 329,248,915 tons of CO_2 equivalents on average, accounting for 63.83% of all host countries' total annual emission reductions. The projects have issued credits worth 419,457,044 tons of CO_2 equivalents, 57.91% of the world's total bringing over 3 billion US dollars to China. Given the current global supply and demand gap of 45,266,854 tons of CO_2 equivalents, the registered projects may generate more than 1 billion US dollars in annual earnings presuming half of them have their credits issued. However, Chinese CDM projects differ significantly in terms of type and among regions due to varied resource access and economic development levels.[1]

B *Challenges to the Chinese CDM Market*

The future, full of opportunities is also rife with challenges. On the domestic level, some projects will be stripped of their novelty as new policies and regulations related to energy conservation and emissions reduction will be unveiled in a faster manner during the 12th Five-Year Plan period. With the promotion of energy conservation, emissions reduction, the elimination of obsolete forms of production, and the strengthening of renewable energy development, the remaining space for CDM project development will narrow. Energy conservation and emissions reduction projects often relate to technological renovation in enterprises, which implies lower costs rather than higher earnings. Therefore, the acceleration of energy conservation and emissions reduction technological advancement will both influence the novelty and reduce the space for CDM project development.

On the global level, international negotiations are hobbling forward. The Kyoto Protocol commitment period only covers 2008 to 2012 while some of the Protocol clauses remain disputed among UNFCCC parties. It remains to be seen whether the Kyoto Protocol will continue after 2012; global efforts against climate change as well as CDMs face a foreboding future, discouraging all project developers. The total value of the global carbon trading market is estimated at approximately 141.9 billion US dollars, down by 1.25% YoY. Due to the temporary lack of post-2012 institutional arrangements, the total value of primary CDM markets have been shrinking for three consecutive years (see Figure 6.1).

---

1 Source: cdm.ccchina.gov.cn (sponsored by Department of Climate Change, NDRC).

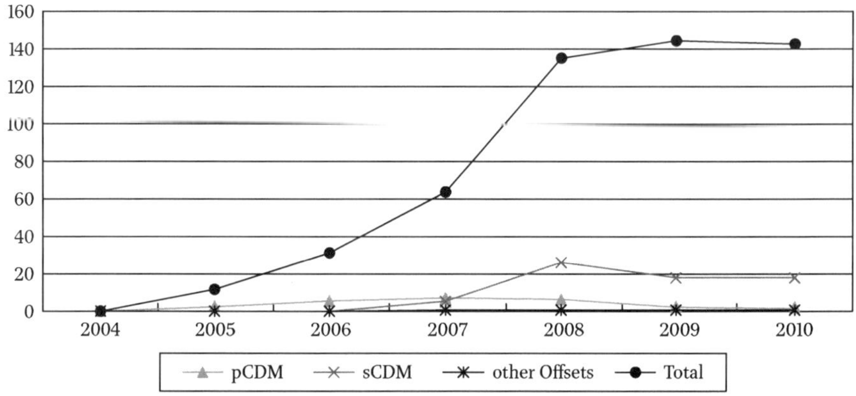

FIGURE 6.1 *Trends of total global carbon market value and major allowance market value (in billion US dollars)*
SOURCE: WORLD BANK, POINT CARBON, ECOSYSTEM MARKETPLACE, UNEP RISÆE.

Compared with the total primary CDM market value from 2006 to 2008 of 5.5–7.5 billion US dollars, the same value in 2010 is only 1.5 billion US dollars, down by 44.44% YoY, which implies a downturn in project credits issued. As a result, the formation of a unified global carbon market is suffering negative effects.

The UN CDM EB has worked out some new rules for project types, regional distribution, and working procedures, seeking to categorize developing countries into different market mechanism implementation levels (see Table 6.1). Such a change is already reflected in the policy arrangement for the post-2012 allowance input, which may produce negative effect on the development of China's carbon market.

TABLE 6.1 *The double-level system for CDM project development*[2]

	Host countries of CDM projects
Non-LDC	China, Vietnam, India, Guatemala, South Korea, Chile, the Philippines
LDC	Cambodia, Madagascar, Rwanda, Uganda, Tanzania

2 Hongbo Chen, 2009, *The Kyoto Flexible Mechanism and the Global Carbon Market. Annual Report on Climate Change Actions: the Road to Copenhagen*, pp. 242–262.

II Policies for the Construction of China's Domestic Carbon Trading Market

Mandatory requirements and administrative approaches, which have been used excessively in China to tackle climate change, have gradually exposed their limitations. During the 11th Five-Year Plan period (2005–2010), China used climate change mitigation as an opportunity to promote the transformation of economic development patterns and economic structures. Energy conservation, emissions reduction, and the development of a green and low-carbon sector have also been recognized as domestic requirements for China's sustainable development. Remarkable results were obtained through a series of policies and actions. For instance, energy consumption per unit of GDP decreased by 19.1% during this period, with 630 million tons of standard coal going unburned and CO_2 emissions reduced by 1.5 billion tons. Chemical oxygen demand (COD) and SO_2 emissions also declined by 12.45% and 14.29% respectively (NPC, 2011).[3] These achievements can be attributed to all central and local governments, which placed great emphasis on energy conservation, emissions reduction, and climate change actions, and further provided critical organizational security for the realization of 11th Five-Year Plan tasks by adopting the chief executive responsibility as well as aggressive administrative approaches. However, these achievements depend on mandatory requirements and administrative approaches such as the elimination of obsolete forms of production, the compulsory shuttering of inefficient power plants, steel factories, and cement plants, and the provision of enormous subsidies. These mandatory requirements and administrative approaches have high economic and social costs, such that sustainability has gradually emerged as a problem in and of itself. In 2010, in pursuit of achieving the energy conservation goal in the 11th Five-Year Plan, some local governments imposed consumption power cuts that seriously impacted corporate production and people's daily lives, drawing wide criticism. It is hence important to rather rely on the market mechanism, which can play a fundamental role in resource allocation, and resort more to macro control over the national economy only in order to set up energy conservation, emissions reduction, and climate change mitigation long-term mechanisms.

The Chinese government has now formally included carbon trading schemes in its most important official documents and national plans. In November 2009, the central government announced explicit targets for

3 Jiabao Wen, *The Report on the Work of the Government*, March 5, 2011, at the Fourth Session of the 11th National People's Congress.

dealing with climate change by 2020, namely that CO_2 emissions per unit of GDP should be reduced by 40–45% from 2005 levels, and the share of non-fossil fuels in primary energy consumption should reach 15%. At the Fifth Plenary Session of the Seventeenth CCP Central Committee, China revealed target for greenhouse gas (GHG) emissions control actions and included the targets into its mid-to-long-term plan on national economic and social development, taking proactive measures against global climate change. China has used GHG emission trading as an indispensable market based method to reduce emissions (CPCCC, 2010).[4] In October 2010, the "Decision of the State Council on Accelerating the Cultivation and Development of Strategic Emerging Industries" (the "Decision" hereinafter) proposed that China establish and improve its major pollutants and GHG emissions transaction system. The NDRC released an interpretation of the "Decision," reiterating the central government's intention of establishing a system for pollutant and GHG emissions trading (CPG, 2010).[5]

In March 2011, the 12th Five Year Plan (2011–2015) on National Economic and Social Development was approved by the National People's Congress (NPC) of China. A series of binding goals covering energy consumption per unit of GDP, CO_2 emissions, COD, SO_2 emissions, NH_3-N emissions, NO_x emissions, forest coverage, and the share of non-fossil fuels in primary energy consumption were also announced, which may directly or indirectly facilitate the fulfillment of GHG emissions reduction goals. The plan's goals form a national economic development system including climate change actions, energy conservation activities, and strict protection for the ecological environment. Energy consumption per unit of GDP, CO_2 emissions per unit of GDP, NO_x emissions, and SO_2 emissions are to be reduced by 16%, 17%, 10% and 8% respectively (NPC, 2011).[6] The Plan also includes measures to establish and improve the GHG emissions measurement and verification system and gradually establish a trading market for carbon credits, as well as an increase to the forest carbon sink. This is the first time that the Chinese government has formally proposed establishing a domestic carbon trading market, indicating penetration into governmental working procedures. The NDRC, along with

4 *The Advice from the Central Committee of CPC on the Development of the 12th Five Year Plan on National Economic and Social Development*, which was passed on Oct. 18, 2010 at the 5th Plenary Session of the 17th Central Committee of the CPC.

5 *The State Council's Decision on Accelerating the Cultivation and Development of Strategic Emerging Industries*, NDRC (2010) No. 32, Oct. 18, 2010.

6 The 4th Session of the 11th National People's Congress, *Outline of the People's Republic of China's 12th Five-year Plan on National Economic and Social Development*, People's Daily, Mar. 17, 2011.

other ministries and committees, have set about designing the carbon trading scheme and undertaking other related fundamental work. Based on the index of energy consumption per unit of GDP mentioned in the 12th Five-Year Plan, the NDRC has worked out implementation proposals for energy conservation, emissions reduction, and GHG emissions control, which seek to develop this work as well as that of carbon intensity abatement by adopting a multi-step pricing mechanism for residential electricity and water use and a metering and charging system for heating, reinforcing fluctuating electricity prices and punitive electricity prices. The NDRC has further directed financial institutions to provide more credit loans for energy conservation, emissions reduction, and low-carbon projects, established a green rating system for banks, accelerated implementation of standards on energy conservation and environmental protection system and the Leader Standard system, and expanded the contents and scope of low-carbon pilots appropriately. Specific measures include:[7] (1) Scientifically and reasonably defining local energy conservation, emissions reduction, and carbon intensity alleviation targets for each region, improving the measurement, monitoring, and verification system, evaluating and auditing the fulfillment of local energy conservation, emissions reduction, and carbon intensity alleviation target, and including the results of such audits into the governmental performance management system as part of an accountability system; (2) carrying out carbon sink and afforestation pilot projects and actively promoting CDM-based afforestation and carbon sink projects to facilitate sound and orderly development of carbon sink forestry; (3) refining laws and regulations on energy conservation and emissions reduction as well as on climate change mitigation; (4) promoting market based mechanisms for energy conservation and emissions reduction and launching more pilot projects of paid use and trading of major pollutant emission allowances in order to gradually advance the establishment of a carbon trading market and to construct and operate a pollution control facility; (5) accelerating establishment of an energy conservation and environmental protection standards system and establishing the Leader Standard system to realize a faster increase in the energy efficiency of energy-consuming products; (6) attempting to establish an identification and certification system for low-carbon products; (7) setting up a GHG emissions measurement and verification system; and (8) strengthening supervision and inspection of energy conservation and emissions reduction work and encouraging all people to save energy, reduce emissions, and fight against climate change, and further clarifying details relevant to key

7 Zhenhua Xie, 2011, "Actively Combating Global Climate Change: An Unshakable Strategic Choice for China," *China Today*, Sept., 2011. http://www.chinatoday.com.cn.

indicators selected from national economic production and life in order to make them more practical and instructive.

The development of China's domestic carbon market has the potential to achieve remarkable progress during the 12th Five-Year Plan period. The NDRC is researching and developing policies on the construction and refinement of carbon market infrastructure, including climate change legislation and a GHG emissions measurement and verification system. The "Management Measures on Voluntary GHG Emissions Reduction Trading Activities in China (Temporary)" (the "Management Measures" hereinafter), which is undergoing development, is already drafted and will soon be open for comments before the approval procedure. The "Management Measures" are an obvious step for domestic voluntary emissions reduction trade. They set up the fundamental registration system for voluntary transactions, define the content and process of voluntary emissions reduction trading, and organize, standardize, and systematize transactions. The Chinese government plans to carve a path towards a mandatory market and to accumulate experience of carbon market regulation through the operation of a voluntary carbon trading market. It will also explore and develop a diversified investment and financing mechanism for climate change actions by carrying out regional carbon exchange pilot projects and then progressively enlarging the exchange to accumulate experience of establishing a national carbon market. It will require confirmation of the qualification of emissions reduction certifiers and will develop uniform standards on certification. More importantly, it will help identify and guide market demands.

III Pilot Actions for Construction of the Domestic Market in China

In accordance with the requirements of its 12th Five-Year Plan, China will build its domestic carbon market comprehensively and holistically and will offer strong support for the development of emissions reducing industries. Ongoing actions on regional or industrial levels mainly include: (1) setting up a domestic system for voluntary emissions reduction; (2) promoting regional pilot projects in five provinces and eight cities; (3) establishing the China Green Carbon Foundation; and (4) voluntary carbon neutral actions by enterprises and institutions.

A Establishment of China's Voluntary Emissions Reductions Mechanism

As a developing country, China is not obligated to undertake mandatory emissions reductions and limitations. However, climate change mitigation is a common obligation shared by all humans. As a responsible country with a large population, China has always prioritized climate change issues, and has elevated actively mitigating climate change to one of its major strategies and permanent policies for economic and social development. It has nevertheless stood by the principle of common but differentiated responsibility and has made considerable efforts and attempts, for example, in voluntary GHG emission reduction.

Guided by national policies and driven by the international conditions, the China Beijing Environment Exchange (CBEEX) and BlueNext jointly developed China's first domestic voluntary carbon standard, the Panda Standard, in December 2009. This standard is meant to match up with China's national circumstances and to be compatible with international rules on voluntary emissions reduction with regard to certification and registration standards. China formally presented the public test version of the Panda Standard at COP 15.

The Panda Standard will facilitate the development of China's voluntary carbon trading market, lending the nascent but expanding market transparency and credibility, and it will fulfill the Chinese government's poverty alleviation objective by encouraging investment in China's rural economy. The Panda Standard will contribute to the Chinese government's efforts to reduce carbon intensity, help develop domestic voluntary carbon trading capacity, and promote agriculture and forestry offset projects with considerable poverty alleviation benefits.

Carbon exchanges have emerged in many provinces and cities in China in recent years. Founding environmental exchanges has proved very popular once the CBEEX, the Shanghai Environment and Energy Exchange, and the Tianjin Climate Exchange were established. Since 2009, exchanges have also been set up in Wuhan, Hangzhou, Kunming, and Dalian as well as in Anhui, Guizhou, Hebei, and Shanxi, which may boost the development of voluntary carbon trading to some extent. Problems have also emerged, such as excessive and irregular exchange management and a lack of service demand.

The exchanges in Beijing, Shanghai, and Tianjin have all been exploring a carbon trading scheme for voluntary emissions reduction. For instance, the Shanghai Environment and Energy Exchange set up a carbon offset platform to support the green World Expo; the Tianjin Climate Exchange initiated a joint corporate action for voluntary emissions reduction; and the CBEEX launched the China Low Carbon Index. China's first trading platform for voluntary carbon credit offsets, called the SEEEX Online Trading Platform, was founded by the Shanghai Environment and Energy Exchange on April 27, 2010, and 526 transactions were completed on it in the first month. The platform has established technical systems, including remote transaction management, immediate quotation, online delivery, and certified standards, as well as a registration and accounting system. Along with further improvements to the trading system and mechanism, the platform will be equipped with the same carbon trading technology used by international institutions.

B *Low-Carbon Pilot Locations in China*

On July 8, 2010, the NDRC issued a notice on "Initiating Low-Carbon Pilot Projects in Provinces and Cities," stating that the five provinces of Guangdong, Liaoning, Hubei, Shaanxi, and Yunnan as well as the eight cities of Tianjin, Chongqing, Shenzhen, Xiamen, Hangzhou, Nanchang, Guiyang, and Baoding have been selected for low-carbon policy pilots.

The NDRC has issued a notice requiring the pilot regions to include comprehensive climate change mitigation in their respective regional 12th Five-Year Plans and to clearly define action targets, key tasks, and concrete measures for controlling their respective local GHG emissions. These cities and provinces are also obligated to reduce their carbon intensity and to explore low-carbon and green development options relevant to their respective local characteristics. They are required to draft corresponding policies supporting low-carbon and green development and to employ a target and task management system to control GHG emissions. The pilot regions will use low-carbon technologies to transform and upgrade traditional industries and will develop low-carbon buildings and transportation alternatives. The regions are encouraged to accelerate establishment of an industrial system with low-carbon emissions and a GHG emissions data, statistics, and management system. Similarly they should actively advocate low-carbon and green lifestyles and consumption patterns. This year, carbon emissions trading has been included as a supplement, encouraging and helping these regions to carry out regional emissions trading pilots by setting caps over local emissions. The NDRC recently convened a number of seminars on the establishment of carbon trading markets in pilot cities and provinces which are meant to kick start such projects.

In accordance with the state's policy, the work on Guangdong Province's pilot project has progressed rapidly. This province is at the forefront of Chinese reform and liberalization and possesses a geographical advantage for facilitating cooperation with Hong Kong (HK) and constructing the SZ (Shenzhen)—HK conurbation, as well as establishing an Asia-wide carbon exchange platform mechanism and system. In order to offer a positive external environment in the process of setting up a carbon credits trading system, the provincial Guangdong government took proactive steps to provide a "green" trading channel for the projects related to the exchange platform construction, to expand the mediums of international communication and exchange, and to disseminate information, raise awareness, and involve the public.

China's low-carbon pilot province and city actions help motivate and engage related forces and accumulate experience distributing work among different regions and industries. They have played a pioneering role and provided many lessons in the development of China's general policies on its domestic carbon market, making a critical case for promoting China's GHG emissions control targets. In 2011, the Institute for Urban and Environmental Studies and CASS worked with the NDRC to create the "low-carbon city evaluation indicator system" with more than 100 indicators, which is expected to facilitate the shift from regional voluntary pilots into mandatory indicator examination and promote low-carbon examination in cities across China.

C *The China Green Carbon Foundation*

The China Green Carbon Foundation was founded on August 30, 2010 succeeding the Green Carbon Foundation established on July 20, 2007 under the China Afforestation Fund. It is the first nationwide, publicly-funded foundation dedicated to combating climate change by increasing carbon sinks. The mission of the China Green Carbon Foundation is to promote activities combating climate change in fields including afforestation, forest management, deforestation and other activities associated with increasing carbon sinks and reducing emissions. It aims to spread knowledge to strengthen public consciousness of how to combat climate change and to support and refine the Forest Effect Compensation Mechanism in China. A brand new operations model was used, in which enterprises or individuals donate to the China Green Carbon Foundation for activities such as afforestation, forest management, etc., and the carbon dioxide absorbed by trees from the forest funded by these enterprises will be credited to their accounts and published on the internet. This projects increase job opportunities for farmers, and raise their income and standard of living by giving them opportunities to participate directly in afforestation activities and forest management, in accordance with the principle that "industry supports agriculture, and the city supports rural areas."

Since its establishment, the China Green Carbon Foundation has collected an endowment of 9.68 million RMB and established five specialized funds in Beijing, Shanxi Province, Zhejiang Province, Daxing region, and Wenzhou County. It has created afforestation pilot projects in nine provinces or cities through partnership with the State Forestry Administration, such that the afforested area has reached 8000 hectares. The first group of 15 individual donors for afforestation has been established in Yan'an City, Shaanxi Province, Jianggangshan City, Jiangxi Province, Duolun County, the Inner Mongolia Autonomous Region, Tengchong County, Yunnan Province, etc., thus inviting public participation in carbon offsetting, carbon footprint elimination, and low-carbon lifestyle.

The foundation has built a four-point benefit platform covering enterprises and the public which includes "storing carbon credits, commercial social responsibility, raising farmers' income, and improving the ecological environment," achieved by forest carbon sinks. The China Green Carbon Foundation guarantees that every ton of carbon offset credit will correspond to a forest plot, which will not only absorb the carbon but also alleviate poverty, increase farmers' income, protect biodiversity and improve the ecological environment. The amount of carbon offset is to be made public and transparent by means of online publication. Increasing carbon sinks is so simple that everyone can participate, thus creating positive social benefits.

D *Carbon Neutral Actions*

In January 2010, the China Carbon Neutral Alliance was officially launched by CBEEX in Beijing. In following with the aims of Green Earth, Sustainable and Harmonious Development through Carbon Neutral, the Alliance aims to provide comprehensive carbon neutral services for enterprises, institutions, and organizations, and to lead in carrying out corporate social responsibility activities and supporting the national strategy on sustainable development.

Launched by CBEEX in association with a large number of professional institutions, the Chinese Enterprises Voluntary Emission Reductions Billboard was presented on June 16th, 2011, alongside 40 other institutions. The institutions include both Chinese and foreign listed companies such as Baidu, Air China, China Everbright Bank, China Merchant Bank, and SocGen. These entities purchased voluntary emission reductions (VERs) to offset GHG emissions generated during their operations or activities, generating a reduction of 210 thousand tons of carbon dioxide equivalents in total.

Among the listed enterprises, China Everbright Bank has made great strides in environmental improvement and corporate social responsibility through various means over the past year. In 2010, the company purchased VERs to offset the carbon dioxide emissions generated in 2009 by the operations of its headquarters and 33 branches. China Everbright Bank is hence China's first carbon neutral bank. It also actively promoted the finances of low-carbon businesses by means of a green credit mechanism, modular financing, and the creation of carbon financing products, turning the primary objective of the financial industry to green and sustainable development.

Besides China Everbright Bank, Air China offered with the first green flight in China, and SocGen issued the first low-carbon credit card—the China Low-Carbon credit card. Baidu became the first internet company to try and offset carbon emissions by purchasing carbon credits.

China has established a preliminary carbon neutral service platform which offers carbon management and carbon neutral services including carbon footprint accounting and verification, carbon neutral trading and certification, as well as carbon assets management and consulting.

China has attached great importance to the construction of a low-carbon industry and system at a national strategic level and has taken many measures and actions that seek to ultimately culminate in a domestic carbon trading market. The construction of a low-carbon industry and system provides a transaction platform for carbon trading, as well as a foundation from which the Chinese government can make relevant policies and rules. The development of China's domestic carbon market will be an effective way to strengthen Chinese companies' capability and knowledge of carbon trading. It will help reduce transaction costs for both buyers and sellers and inject greater liquidity into the global carbon market. The efforts that China is making at present in the field of low-carbon development can provide practical experience for policy formulation dealing with the nationwide carbon market and is an irreplaceable and inevitable step in the development of the future carbon trading scheme.

IV Prospect for China's Domestic Carbon Market

The emissions trading market mechanism is marred by complicated details and its establishment adds many more layers of complexity, so it will be established in a stepwise process beginning with voluntary trading markets and moving on to regional carbon exchange pilot projects, and finally the establishment and improvement of a nationwide carbon market. China's domestic carbon market is expected to follow an evolutionary track starting at voluntary and shifting to mandatory and starting regionally and shifting to a nationwide market.

At first, China will establish a voluntary carbon trading registration system based on the "Management Measures on Voluntary GHG Emissions Reduction Trading Activities in China (Temporary)" to standardize the voluntary GHG emissions trading activities in China, to ensure openness, fairness, and transparency of the voluntary market, to stimulate entrepreneurial enthusiasm for participation in climate change mitigation actions, to boost the appeal of demand-side GHG emissions reduction credits, and to resolve the lack of a domestic voluntary carbon market credit system. The central government will encourage and help qualified regions and industries with attempts in carbon trading, looking to launch carbon trading pilot projects in some industries and provinces within five years. Although the "Management Measures" have not yet defined the standards and pricing rules for voluntary emission reductions, they have specified the trading products, trading location, new methodology application procedures, and DOE qualification procedures for voluntary emission reductions trading, moving the processes of emission reductions certification, project registration, emission reductions issuance, etc. under the control and jurisdiction of the respectively relevant authorities. The second step will be to launch the regional carbon trading pilot projects, which will accumulate experience for a nationwide carbon market, explore a diversified investment and financing mechanism for climate change mitigation activities, and proactively guide foreign investment to R&D of low-carbon technologies and low-carbon industrial development.[8] Third, using the experience gained form the pilot projects. China will gradually enlarge the scope of trading and set up a nationwide carbon market in a scheduled and stepwise manner.

The development of China's domestic market relies on the progress of market infrastructure construction. There is no exact schedule for the construction of the Chinese carbon market at present, but it is certain that during the 12th Five-Year Plan period, remarkable progress will be achieved. The operation of the voluntary trading market will actively explore solutions for a mandatory market and will accumulate experience for the official regulations, the link to foreign markets and the balancing of pressures in the international market. Development of CDM projects will continue within a certain scope while new bilateral and multilateral trading mechanisms are likely to be launched, thus laying the foundation for the link of liquid assets between China's domestic

8 In 2010, five provinces including Guangdong, Liaoning, Hubei, Shaanxi, and Yunnan as well as eight cities including Tianjin, Chongqing, Shenzhen, Xiamen, Hangzhou, Nanchang, Guiyang, and Baoding started to carry out low-carbon development pilot projects with carbon trading as their key project item. These pilot regions were encouraged and supported launching regional carbon trading by setting a cap on local emissions. Located in developed coastal areas of eastern China as well as in central and western China, these pilot regions differ from each other in terms of natural resource reserves, natural conditions, economic level and industrial structure, which will help accumulate instructive experience for the establishment of local carbon markets and the promotion of low-carbon and green development in different geographic areas.

carbon market and the international carbon market.[9] Due to the enormous effect of the economy and the multistage nature of its development, it is unquestionably necessary to persist in inclusive growth, to set up systems and mechanisms and establish China's own trading platforms as well as incorporate representatives from all levels of the trading regime in order to ultimately achieve an internationally compatible system. It is likewise necessary to propel carbon trading and to accumulate experience and build up domestic carbon resource reserves. In this way, China will convert itself from a mere supplier of GHG emission reduction credits into an advanced participant in global and regional carbon markets.

(This article was originally published in Chinese in 2012.)

9 According to the Cancun Agreement, the future market mechanism for the joint efforts of industrial and underdeveloped countries against climate change will still originate from the CDM while improvement will be made with respect to the mechanism and rules, process, transparency and working efficiency, and the problem of lengthy project approval is to be solved.

CHAPTER 7

Analysis of Synergistic Effects of Low-Carbon Actions and Climate Change Adaptive Measures

Wang Wenjun and Zheng Yan

Abstract

The collaborative actions of mitigation and adaptation have stirred new discussions in the research of addressing climate change. This paper summarizes the literature and practices of collaborative management of mitigating and adaptive actions. The study shows that it is feasible to manage the actions of mitigation and adaptation collaboratively. Using key elements of synergistic management, this paper conducts an in-depth analysis on the synergistic effects produced by three types of adaptation and mitigation actions in the energy sector. The research found that about half of the actions could produce synergistic effects. Finally, this paper studies the possibility of collaborative management in addressing climate change in China's Guangdong Province, and gives suggestions based on the analysis.

Keywords

low-carbon development – adaptation actions – synergistic effect

Reducing greenhouse gas emissions (GHG) and adapting to climate change are two crucial tasks for minimizing climate change risks. (IPCC AR4, Working Group I & III, 2007). Even if the most drastic emissions reduction measures are adopted, we may not avoid the negative impacts of global climate change on human society, so that actions of adapting to climate change are imperative.

* Wang Wenjun, Guangzhou Institute of Energy Conversion, Chinese Academy of Sciences, email: wangwj@ms.giec.ac.cn.

Zheng Yan, Institute of Urban & Environmental Studies, Chinese Academy of Social Sciences.
This paper is funded by the major research program of National Natural Science Foundation of China (No. 70933005).

 | DOI 10.1163/9789004274648_008

If reductions and mitigation fail to be implemented, the dramatic effects of climate change would sooner or later undermine human efforts at sustainable development. The effective management of reduction and mitigation measures alongside increasing local adaptation capability by means of mitigation and incorporating low-carbon measures into adaptation actions should undoubtedly be the ultimate goal of climate policies for most regions of China. As one of the largest GHG emitters in the word, China is faced with the herculean tasks of reducing its GHG emissions, while many regions struggle with vulnerable economic, natural, and geographical conditions, and simultaneously adapting to the changing climate system. It is essential to manage actions for mitigating and adapting to climate change in China collaboratively.

I Research Background and Overview: Collaborative Management of Low Carbon Measures and Adaptation Actions

Reducing GHG emissions, as an important component of low-carbon development projects, is closely linked to projects in human adaptation to climate change. The spatial and temporal patterns of GHG emissions reduction interact with those of adaptation, and the implications and effects of mitigation overlap with those of adaptation (Gitay et al., 2001; Vellinga et al., 2001; Cohen et al., 2001). In the process of addressing climate change, policy makers should take these points of interaction into consideration, identifying and effectively managing the synergistic effects from those actions. Low-carbon development can thus be efficiently achieved with low costs and alongside sustainable socioeconomic development.

A *Research Background*

Before the concept of collaborative management of mitigation and adaptation actions was proposed by the IPCC, mitigation and adaptation were treated as two independent actions and consequently policy makers tended to develop two separate sets of action plans to address them. Mitigation-adaptation interactions were not taken into consideration during policy design. Thanks to ongoing research on global climate change, many scholars have found that while many measures were designed to adjust human societies or industries to prepare for climate change, these measures were likely to increase GHG emissions and undermine low-carbon development efforts. On the other hand, some low-carbon measures, although economically and technically feasible, might increase the ecosystems' vulnerability and as a result reduce many local areas' resiliency against climate change effects. Given this paradox, some scholars began observing the relationship between adaptation and sustainable development projects (Metz, 2000; Beg et al., 2002; Markandya and Halsnæs, 2002; Klein and Smith, 2003; Lin, 2007). The studies so far have no directly researched collaborative management of mitigation and adaptation projects.

The concept of synergistic effects derived from the collaborative management of mitigation and adaptation actions was first articulated by the IPCC Third Assessment Report (IPCC TAR, 2001a) and echoed by some scholars (Moomaw et al., 2001; GAIM Task Force, 2002; Clark et al., 2004). As the mitigation-adaptation relationship has not yet been clarified, current research focuses on collaborative management concepts as well as relative research methodologies, most of which are qualitatively oriented. After six years of exploration and research, the world climactic community formally put the collaborative management of mitigation and adaptation actions on the research agenda. An additional chapter was added to the IPCC Fourth Assessment Report (IPCC AR4, Chapter 18, 2007a), which aims to present and overview and analysis of the collaborative management of mitigation and adaptation actions, and calls for climate researchers to conduct quantitative studies on the synergistic effects of mitigation-adaptation actions. The chapter indicates great promise in quantitative studies on the topic.

Synergistic effects can best be understood as the results produced by the collaboration and operation of separate sub-systems within a general system for a common goal. Synergy can be simple expressed with the formula "1+ 1> 2."[1] According to this logic, we may argue that if human's response to climate change is viewed as a singular system, then the subcomponent measures aimed at reducing GHGs in the atmosphere that also have the potential to alleviate the negative impacts of climate change on human beings and natural environment, or the subcomponent measures aimed at reducing the negative impacts from climate risks that may also decrease GHGs in the atmosphere can be said to have a synergistic effect. Purposefully employing collaborative management of mitigation and adaptation strategies, known as synergistic management, can achieve greater synergistic effects.[2]

1 Bai Liehu, "Theory of synergy and theory of synergy of management," *Social Sciences of Gansu*. Issue 5, 2007, 228–230.

2 Wang Wenjun and Zhao Daiqing, "Studies on the development of mitigation and adaptation synergies: the case of Guangdong," *China Population Resources and Environment*. Issue 6, 2011, 89–94.

B *Literature Review: How to Identify the Synergistic Effects of Mitigation and Adaptation Actions*

Since 2007, many scholars have begun to explore the synergistic relationship between mitigation and adaptation projects. The research focused on three aspects: determining preexisting synergistic relationships in mitigation and adaptation actions; analyzing the feasibility of collaborative management; and optimizing synergistic effects.

Changing climate systems impact human societies in different ways. Mitigation and adaptation could be complementary in some respects but conflict in others. Identifying which actions can be managed collaboratively in order to reap synergistic effects in necessary. Research of this aspect is regrettably underdeveloped. Taylor (2006) listed some possible mitigation-adaptation interaction scenarios: mitigation measures impact adaptation targets, adaptation actions impact mitigation goals, no interaction between mitigation and adaptation actions, and mutual impacts between mitigation and adaptation. Strong interactions between mitigation and adaptation actions have been fully unearthed by this type of research, but or not taken into account during policy formulation (Boehm et al., 2004). Peters (2001) makes a strong argument that the impacts of mitigation actions on energy, transportation, residence, commerce, and industries adaptation targets were grossly underestimated.

1 Feasibility of Synergistic Effect Management

There are two different arguments regarding the feasibility of mitigation-adaptation synergistic management. Some optimistic scholars hold that synergies have the potential to generate multiplier effects as a result of improved physical and institutional design (Venema and Cisse, 2004; Goklany, 2007; Biesbroek et al., 2009). Sirower (1997) presents a dissenting argument of the "synergy trap." Other scholars argue that although synergistic management of mitigation-adaptation strategies should be implemented, it remains uncertain if such efforts are cost effective (Klein et al., 2005; Sovacool and Brown, 2009). After a close examination of the feasibility of mitigation-adaptation actions synergistic management, British scholar Tyndall Center argued that since both mitigation and adaptation actions were driven by a common set of factors, synergistic effects could appear at the local level (J.T. Richard et al., 2003). The scholars of the Chinese Academy of Social Sciences (CASS) developed the concept of incremental and developmental adaptations. They argue that the synergistic effects derived from different types of adaptation actions may take place in different areas (Pan and Zheng, 2009).

2 Optimization of Synergistic Effects

There is no clear optimal scheme for managing mitigation and adaptation projects. As local geographical conditions, policies, institutions, and degree of management vary across time and among various impact factors, it is rather difficult to find an optimal solution (IPCC TAR, 2001). Some studies have shown that seeking optimal systems is unnecessary. They propose concentrating on a range of synergistic effects and profiting from multiplier effects or lowered costs (Wilbanks, 2007). Other scholars believe that the effects of collaboration depended upon the ability to properly manage projects (Yohe, 2001; Adger et al., 2003; Adger and Vincent, 2005; Brooks et al., 2005).

C *Progress and Practice in China and Abroad*

Purposeful management of mitigation and adaptation actions aimed at synergistic effects is scarce due to the different pathways of top-down tendency of mitigation actions and bottom-up tendency of adaptation actions, such that synergy cannot emerge spontaneously but only through policy designing. The goal of current climate policies is to reduce GHG emissions, with no real focus on adaptation. Mitigation and adaptation actions are not intertwined systematically leaving room only for few and fragmented synergistic effects. For example, the Noel Kempff Mercado Climate Action Project in Bolivia and a regional development scheme in UK (ODPM, 2005) attempted to collaboratively manage mitigation and adaptation projects. The Noel Kempff Mercado Climate Action Project in Bolivia achieved three goals through well oriented forest planning: capturing carbon dioxide, protecting ecosystem biodiversity, and promoting the sustainable development of the local economy. Germany has improved the eco-function of its rivers and its flood prevention capability by demolishing cement embankments and widening its urban water systems. Carbon emissions are reduced due to the cement embankments demolition. Some European cities have employed stereoscopic greening technologies in order to build gardens and lawns on the top of buildings. These gardens and lawns can reduce urban heat island effects and increase carbon absorption, as well as hold rainwater reducing urban flooding disasters. These cities ability to adapt to climate change will be improved as a result.

II Analysis of Key Collaborative Arenas and Synergistic Effects

The international climate treaty states that effective efforts to tackle climate change require the involvement of all major GHG emitters. Specific adaptive measures are largely dependent upon the type of climate risks a region is suffering from and the level of vulnerability a particular region, industry, or a particular group of people have with regard to climate risks. Adaptations are also subject to the influence of government policies, including through policy administration at the local level, such as communities, businesses or individuals.[3] Can mitigation and adaptation produce synergistic effects? Is it possible to manage synergy? Research into these two issues should be approached from both the state and local level. This paper explores the mitigation and adaptation projects in certain key fields as defined by China's national program on climate change and it defines the suitable scope for synergy. This paper continues with an analysis of mitigation and adaptation actions' synergistic effects within energy industries. This paper first attempts to analyze and classify mitigation and adaptation actions according to their interactions.

A *Analysis of the Interactions of Mitigation and Adaptation Actions*

Mitigation and adaptation actions may produce effects irrelevant to their original goal, which are known as externalities, and may have either a positive or negative impact. An externality can be categorized as arriving from mitigation based actions or adaptation based actions, and as positive or negative externalities.

1 Emissions Reduction Externalities from Adaptation Projects

- Adaptation initiative cause an increase in emissions—adaptation actions' negative externalities affecting mitigation

Humanity's endeavor to adapt to climate change might increase GHG emissions, which is usually called adaptation activities emissions. For example, the usage of carbon-intensive products, like cement, lime, steel, etc., for fortifying city infrastructure, or building water conservation facilities and breakwaters, has the potential to increase carbon emissions. Cultivating high-yield crop varieties by using fertilizer and pesticides may reduce soil fertility and increase nitrous oxide emissions. Adaptations actions may also generate waste. Without sophisticated waste incineration technology, waste can only be deposited in a

3 R.J.T. Klein, S. Huq, F. Denton, T.E. Downing, R.G. Richels, J.B. Robinson, F.L. Toth, 2007: "Inter-relationships between adaptation and mitigation," in *Climate Change 2007: Impacts, Adaptation and Vulnerability. Contribution of Working Group II to the Fourth Assessment Report of the Intergovernmental Panel on Climate Change*, edited by M.L. Parry, O.F. Canziani, J.P. Palutikof, P.J. van der Linden and C.E. Hanson (Cambridge, UK: Cambridge University Press), 745–777.

landfill or composted, leading to inevitable carbon emissions. Carbon emissions may become particularly serious when governments are developing infrastructure.

- Mitigation effects from adaptation activities—adaptation actions' positive externalities affecting mitigation

The development of technologies that seek to eliminate forest pests and diseases and eliminate forest fire, such as using cold, drought, and bug-resistant tree varieties has the potential to minimize the impact of climate change on biodiversity. These measures may also improve the forest carbon sink capacity. Coastal areas are rich in wind and solar power. Policy makers should encourage businesses to use clean energies to the greatest extent possible when developing infrastructure. For example, developing tidal power, offshore wind power, and solar photovoltaic power may replace some fossil fuel based power plants and help reduce GHG emissions. Similarly, ecological projects, like the protection of wild plants, animals, and natural reserves, the development of forest reserve for biomass energy, and the green wall, will all contribute to the protection of the forest carbon sink. As a result, the land's carbon storage will increase.

2 Adaptation Externalities from Mitigation Actions

- Positive adaptation externalities from mitigation actions

The low-carbon activities designed to reduce GHG emissions may positively impact mitigation efforts in two ways. First, in the near future, improving energy use efficiency can provide strong low-carbon material support to adaptation initiatives. The agricultural and husbandry development could be protected and strengthened by afforestation projects, turning farmland into forests and pastures, constructing a green wall system, etc. This could effectively reduce the risks posed by climate change to agriculture and biodiversity. Second, as GHGs atmospheric presence decreases thanks to mitigation efforts, climatic disasters will become more infrequent, allowing investments for adaptation to be used for other purposes.

- Negative adaptation externalities from mitigation actions

Mitigation efforts seek to transform the current energy usage pattern to cleaner structures. However, traditional hydropower stations inevitably harm the local environment; biomass power generation requires a large supply of plants, whose harvesting may weaken local climate change adaptation capability in some areas.

B *Key Mitigation-Adaptation Interactions Arenas*

According to China's national program on climate change, the key mitigation arenas include energy industries, industrial sectors, agriculture, forestry and the production of daily necessities. Among all these sectors, the energy industry may be the key sector on which China will focus mid-term and long-term mitigation activities, such as nine measures to reduce GHG emissions that have already been developed. Adaptation is to occur predominantly in agriculture, forestry, water resources management, as well as in management of coastal belts and coastal areas. Agriculture and forestry are prime interaction arenas for mitigation and adaptation actions, so that collaborative management of the two efforts in those arenas has a great possibility of producing synergistic effects. Adaptation initiatives may be sorted into three categories: engineering-based adaptation, technology-based adaptation, and institutional adaptation.[4] Among these three types of adaptation, engineering-based adaptation has the closest relation to energy and industry. Engineering-based adaptation may create extra emissions or mitigation effects. Most mitigation activities interact with adaptation efforts within every key arena specified by China's national program on climate change. Figure 7.1 demonstrates the possibility of collaborative management in the fields of energy and agriculture.

Engineering-based adaptation will inherently produce some emissions. For example, engineering projects may strengthen the adaptive capability of the socioeconomic system to climate change, but the process of building these projects may change land use patterns and increase demand for energy-intensive and high carbon products, increasing carbon emissions. If the engineering-based projects could be integrated with low-carbon projects, emissions from engineering-based projects could be reduced by using renewable energy or the projects could lower carbon intensity through improvements to energy efficiency. For example, if the design of dam could take the particular characteristics of a hydropower station into consideration or vice versa, it might decrease the negative externalities caused by the mitigation and adaptation actions.

4 Pan Jiahua and Zheng Yan, "Analytical framework of adapting to climate change and policy implications," *China Population Resources and Environment*, Issue 10, 2010, 1–5.

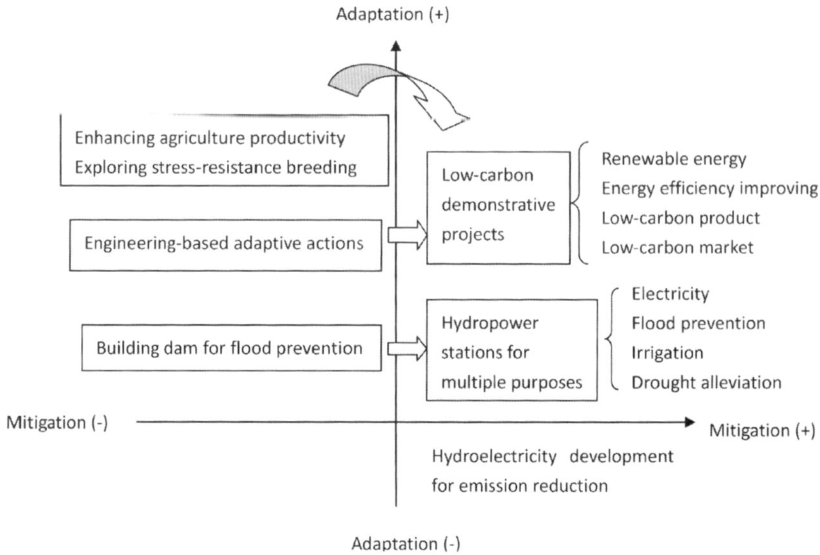

FIGURE 7.1 *The possible points of collaborations between mitigation and adaptation actions*

C *Analysis of the Synergistic Effects of Mitigation and Adaptation Activities*

Among the key arenas and classifications for climate adaptive actions mentioned above, energy related mitigation activities will most closely overlap with engineering-based adaptation actions, industrial and agricultural mitigation overlaps with technical adaptation, and mitigation policies dealing with forestry and residential consumption overlap with institutional adaptation. Many mitigation and adaptation measures exist within socioeconomic system projects, so that it is impossible to describe every mitigation and adaptation pairing that can produce a synergistic effect in this paper. Instead it will try to distinguish mitigation activities in the energy sector that produce a synergistic effect with three adaptation actions.

1 The Limits of Mitigation and Adaptation Activities

There are three major categories of adaptation activities. Engineering adaptation activities mainly include the construction of water conservation facilities, environmental infrastructure, and inter-basin water transfer engineering. Technical adaptations include research and development of new crop varieties, and new technologies that can help ecosystems adapt. Institutional adaptation activities aim to enhance the ability to adapt to climate change by means of policy, legislation, and system construction, such as establishing a carbon tax, providing compensation for watershed ecology, disseminating popular science, etc. (Pan Jiahua, Zheng Yan; 2010).

Emission reduction activities can be divided into four main groups.[5] In the energy sector, the first grouping is the production of low-carbon energy, which includes the development of hydropower, nuclear power, solar photovoltaic, solar thermal power, wind power and other low-carbon or carbonless energy development. Low-carbon energy consumption refers to the consumption behavior and the usage of low-carbon or carbonless energy. For example, the British city Modun requires that all commercial development projects greater than 1000 square meters must have at least 10% of their energy derived from renewable sources.[6] The resulting huge spike in demand for low-carbon energy will accelerate its production and commercial scale. Energy efficiency improvement refers to adopting measures such as energy-saving technological exploration, energy management contracts, implementation of integrated resource planning, demand-side power management, etc., with an objective of reducing GHG emissions using the energy structure. The last group is innovation and construction that addresses the climate change mechanism, including the establishment of a pricing system, which can help achieve a low-carbon energy structure adjustment, building a stable financial investment mechanism, and policies that strengthen energy conservation supervision and inspection.

It is better to study synergistic effects anchoring around the reaction of individual actions. Only thus can the cost-benefit analysis of mitigation and adaptation synergistic effects be accurately reflected. The categories of synergistic actions are separated by the type of measure, which can only determine the magnitude of the synergistic effect.

2 Analysis of Synergistic Effects of Mitigation and Adaptation Activities

Using the categories established above, the following steps are taken to discern whether the given mitigation and adaptation activities may produce synergistic effects.

The first step is to determine whether the two actions have a synergistic effect due to a mitigation action externality, that is, if a mitigation measure will be beneficial to an adaptation goal, the two actions can be said to have a synergistic effect, a "+" is used to identify this effect. Conversely, if a mitigation action

---

5 Extracted from "China's national program on climate change," http://www.ccchina.gov.cn/WebSite/CCChina/UpFile/File189.pdf.

6 Ye Zhuda, "Road to low-carbon development," *Jiangsu City Planning*, 2009(7): 6–10.

has no adaptation externality, a "0" will be used; if the externality is negative, a "–" will be used. See Table 7.1.

The second step is to determine whether the two actions have a synergistic effect due to an adaptation action externality, that is, if an adaptation measure is beneficial to a mitigation goal, the two actions can be said to have a synergistic effect, a "+" used to identify this effect. Other signs follow the pattern in the first step. See Table 7.2.

The third step is a comprehensive assessment of the extent of synergistic effects based on the first two steps. Four possible situations are: strong synergistic effect, which is when both of mitigation and adaptation action positively affect one another, which will be denoted with a (+, +). A (+, 0) or (0, +) denotes a mitigation or adaptation action that is beneficial to its counterpart without the corresponding feedback, which is called weak synergistic effect. Proper policy formulation can ensure that synergy is nevertheless achieved. Whenever a (+, 0), (0, +), or (+, +) appears, it means more detailed classification is needed to determine the detailed possibilities of the synergistic effects. See Table 7.3.

TABLE 7.1 *Mitigation measures' adaptive externalities*

	low-carbon energy production	low-carbon energy consumption	energy efficiency improvement	low-carbon mechanism innovation
adaptaion engineering projects	+	+	+	±
technical adaptation projects	+	0	0	+
systemic adaptation activities	0	+	+	±

TABLE 7.2 *Adaptive actions' mitigative externalities*

	adaptation engineering projects	technical adaptation projects	systemic adaptation activities
low-carbon energy production	+	+	+
low-carbon energy consumption	+	0	–
energy efficiency improvement	0	+	±
low-carbon mechanism innovation	+	+	±

Tables 7.1 and 7.2 reveal that some adaptive measures will be beneficial to the goal of reducing GHG emissions, and vice versa. Confirming the synergistic effects between adaptive actions and mitigation measures requires integrating perspectives and combining Tables 7.1 and 7.2, which is presented in Table 7.3.

TABLE 7.3 *The synergistic effect of mitigation and adaptive actions*

	low-carbon energy production	**low-carbon energy consumption**	**energy efficiency improvement**	**low-carbon mechanism innovation**
adaptation engineering projects	(+, +)	(+, +)	(+, 0)	(±, +)
technical adaptation projects	(+, +)	(0, 0)	(0, +)	(±, +)
systemic adaptation activities	(0, +)	(+, −)	(+, ±)	(±, ±)

Note: the first sign in the pair represents mitigation measures' to adaptive externality and the second the adaptive actions' mitigative externality.

Table 7.3 shows that there are three combinations yielding strong synergistic effects and two combinations yielding weak synergistic effects out of the 12 possible combinations. It is difficult to determine whether or not a synergistic effect exists for the remaining seven combinations. The synergistic combinations fall into three categories: the combination of new adaptive engineering with low-carbon energy supply and demand, which can reduce GHG emissions and climate risk; technical adaptation measures, such as farmland drought relief measures combined with mitigation measures in the energy sector have ranging degrees of synergy; and energy efficiency improvement, which simultaneously saves energy costs and reduces carbon emissions.

It is suggested that the externalities caused by climate change strategy policy be taken into account in order to improve policy priority. Following is a case study analyzing the synergistic effects of mitigation and adaptive actions in Guangdong Province.

III Collaborative Management of Mitigation and Adaptation Actions

Guangdong province is located at the southern end of Eurasia and within range of the South China Sea monsoon area. It is near the sea, with a long coastline of 4114 kilometers it is particularly climate sensitive. During China's 44 major natural disasters, over 40% have taken place in Guangdong and 80% of meteorological disasters have taken place there. With economic structural transformation, Guangdong's disaster areas are gradually shifting from agriculture to second and third tier industries. As the largest energy and power consumption province in China, Guangdong's per capita carbon emissions level is higher than the Chinese average and its per capita energy consumption is over double the national average. In the 12th Five Year Plan period, Guangdong faces the daunting challenge of decreasing unit energy consumption and unit carbon emissions by 18% and 20% respectively compared with 2010. Guangdong must reduce carbon dioxide emissions and adapt to climate change in order to address both its national commitments and its geographical vulnerability.

A *Key Mitigation and Adaptation Actions in Guangdong*

Guangdong's recent carbon emissions from electricity and industry made up over 80% of total emissions. 70% of the energy consumed in Guangdong was produced through coal burning. In 2007, Guangdong had a total installed capacity of 58.8 million kilowatts of energy production capacity, of which thermal power made up 76%.[7] To reduce carbon emissions from energy and industries will be a primary target of future mitigation actions, which will implement low-carbon energy structures. Another implication is that emissions reduction effort will focus on the electricity and industrial sectors.

Guangdong province has a low elevation and higher climate risk. Some studies have verified that most of the Pearl River Delta may be submerged, up to 1153 km2, if the highest tide level rises by 30 cm. Guangzhou, Zhuhai, and Foshan are Guangdong's most threatened cities. According to a climate risks assessment report made by the OECD in 2007, Guangzhou was one of the highest climate risk cities in the world. If adaptation activities have not been implemented, economic damage may reach as high as 56 billion RMB in 2030.

Guangdong's adaptation activities will prioritize the infrastructure of coastal cities and the protective construction of coastal zones according to the list of adaptation activities in the "National Climate Projects" and the nature of damage cause by meteorological disasters. Using the analysis above, specific measures can describe and embody the items listed in the table in the following ways: the construction of water conservation facilities and flood control dams are examples of adaptation engineering projects, as well as seawall

7 Yi Jingwei, "Study on the Transition to Low-Carbon Electricity in Guangdong: Pathway, Policy and Economy," dissertation, University of Science and Technology of China, 2010.

engineering in coastal zones, disaster defense measures and expanding cities' green areas to disaster prevention and reduction. Construction of hydropower, solar power, and wind power projects to promote low-carbon energy production, the compulsory purchase of clean energy to promote low-carbon energy consumption; and enhancing low-carbon education and low-carbon innovation are other ways to match projects with table categories.

Table 7.4 shows that there is a strong synergistic effect between water conservation facility construction, low-carbon energy production, and clean energy consumption. Mitigation measures and adaptive actions are mutually beneficial in climate change education and awareness programs, popular science projects that yield positive externalities of encouraging the use of clean energy and improving energy efficiency, and the city greening project that promote energy savings.

TABLE 7.4 *Synergistic effects of mitigation and adaptation measures in Guangdong*

water conservation facilities	(+, +)	(+, +)	(+, 0)	(+, 0)
seawall engineering	(0, +)	(+, 0)	(+, 0)	(+, 0)
city greening projects, popular science programs	(0, 0)	(0, +)	(0, +)	(+, +)

B *How to Manage the Synergistic Effects of Mitigation and Adaptation Actions*

Studies have found that varied methods should be employed to provoke a synergistic effect dependent upon which mitigation and adaptation actions are interacting. Two kinds of effect will be used as examples to illustrate how to manage strong and weak synergistic effects caused by mitigation projects.

1 Strong Synergistic Effects (Hydropower Station Construction versus Water Conservation Facilities)

Guangdong has about ten hydropower stations, which are scattered among Conghua, Shenzhen, Dongguan, Zhuhai, Guangzhou, Zhanjiang and other cities. Assuming extreme weather events increase, if the hydropower stations are equipped with anti-flood and drought resistance facilities, these stations can irrigate and store water as needed. A synergistic effect can be provoked at the highest level, where water resources are used for power generation, but also for flood storage and drought resistance, which enhances the system's resilience against climate change.

A hydropower station is restricted to water resource distribution. Guangdong's hydropower stations are near maximum capacity, it is impossible to increase the proportion of clean energy by constructing larger hydropower station. Instead, Guangdong is vigorously developing solar energy, offshore wind power, and nuclear power. If policy designers take enhanced adaptive capacities into consideration during planning of low-carbon energy construction, huge synergistic effects can be achieved. By managing the structure of energy consumption and the ratio of low-carbon energy in construction, thereby increasing demand for low-carbon energy, the cost of low-carbon energy production will be decrease significantly, which will provide a greater impetus to scale low-carbon energy production. For example, when a city needs to fortify its infrastructure or rebuild bioswell prevention systems to confront climate risks, solar photovoltaic building renovations offer an alternative way to build and renovate public building, making use of inexhaustible solar power that is also not vulnerable to extreme weather, making it a prime method for adapting to possible power failures following climate disasters. The application of solar energy further increases the proportion of renewable energy in the total energy structure, which works to mitigate climate change.

2 Weak Synergistic Effects (City Greening and Education versus Clean Energy Usage)

City greening measures, which vary depending on the city's general landscape plan, aim not only aim to make communities more beautiful but also to abate negative effects from climatic disasters using design and planning. For example, a park landscape blueprint can be designed with flood storage, irrigation, and bioswell prevention all worked in together, enhancing the city's water storage capability in the event of a rainstorm or a drought. These elaborate landscape parks will provide a cushion and automatically decrease the damage from climatic disasters. Low-carbon awareness raising can encourage more people to use low-carbon energy.

IV Experience Diffusion and Policy Suggestions: Synergistic Policies

In addition to the synergistic effects discussed, many externalities exist in other fields and areas. As these channels addressing climate change are low cost, it is worthwhile exploring and sharing possible synergistic effects of mitigation and adaptation actions.

A *The Goals and Principles of Policy Management of Synergistic Effects*

After identifying the intersection of positive externalities of mitigation and adaptation actions, and formulating comprehensive plans and policies based on those intersections, the goal of collaborative management is to further amplify the positive externalities and increase total social welfare. Different countries and regions find themselves in different stages of development and have varied social benefit functions. For example, the most important social welfare factors in developing and less developed countries are per capita income, literacy, health, and the gap between rich and poor, while environmental quality is one of the most important social welfare factors in developed countries. Therefore, when we work out the synergistic management policy, some key principles must be followed.

The first principle is that of rational decision making. That means using the assessment of climate risk as a scientific basis and considering local climatic risks and carbon emissions structures to establish a matrix of mitigation and adaptation actions, and then employing the Delphy method to analyze possible synergistic effects.

The second principle is employing holistic judgment. Some action pairs may have a strong synergistic effect but may be costly. When making a decision regarding mitigation and adaptation project management in order to realize synergistic effect, more criteria should be included such as cost-benefit analysis, stakeholder surveys, etc., and not just the potential synergistic effect.

The third principle is maintaining consistency in policy. It is necessary consider possible conflicts inherent in departmental cooperation, such as the current mitigation or adaptation policies and synergistic action designs.

B *Key Factors of Synergistic Effect Management*

The socioeconomic system would enjoy a great benefit if mitigation effects are considered in the design of adaptation actions, however, it must be understood that adaptation and mitigation activities are inherently different. This disparity should be respected when engaging in the following activities dealing with synergistic management. These activities are: searching for relationships between actions, finding appropriate technologies collecting greenhouse gas emissions data, predicting future emissions scenarios, and working out the solution, and identifying and coordinating stakeholders, and integrating various management systems into a policy pool to prevent mutual interference.

There are three necessary steps for synergistic management: limiting the scale of synergistic actions, running a cost-benefit analysis for each action pair, comparing the cost of collaborative actions with individual measures, and analyzing the feasibility of every collaborating action pair.

To analyze each collaborative action pair falls squarely in the hand of policy designers. Countries and regions with different carbon structures and climate risks have varied activities addressing climate change respective to their conditions. Some activities are similar, some are different, and that is why analysis of local condition is important and necessary. Effective management that takes local conditions into consideration can get twice the results for half the effort.

C *Policy Suggestions: How to Put Collaborative Actions Addressing Climate Change into Effect*

Mitigation and adaptation actions include technology, mechanism, and behavior selection guided by policy and the possibility of greater efficacy, so that mitigation and adaptation policies can be integrated into a national climate strategy. Collaborative management based on the study of managing climatic actions is an important path of local governments to address climate change.

(1) Mitigation and adaptation actions have an interactive relationship at every level (global, national, regional and individual). The primary precondition for executing collaborative management is to find the synergistic effects of mitigation and adaptation actions. Collaborative management methods should be designed according to the features of actions at different levels to efficiently guide individual and community spontaneous behavior.

(2) Establishing a policy pool of mitigation and adaptation actions, ranking the priority ordering of the actions according to different management objectives, and running a cost-benefit analysis, provides clear collaborative management documents that can be reviewed whenever needed.

(3) The arena best suited for seeking synergistic effects of mitigation and adaptation actions depends policy makers or designers judgment,[8] so that policy makers and designers must be adequately trained to ensure the effectiveness of collaborative management.

(4) Collaborative work should be encouraged in basic data collection. Adding the database of adaptive initiatives to that of GHG emissions can enlarge the functionality of the data system. The larger data pool is conducive to strengthening public health services and disease prevention by recording

8 Thomas J. Wilbanks, Jayant Sathaye, "Integrating mitigation and adaptation as responses to climate change: a synthesis," *Global Change*, 2007: 957–962.

the impact of climate change on human health, thereby improving the system's ability to adapt and forecast climate change.

(5) Collaboration between training and awareness raising should be strengthened. Mitigation and adaptation are both societal responses to climate change, and despite the different objectives and tasks, they are both new to the general public. Few people are aware of mitigation and adaptation initiatives, thus, it is necessary to train and raise awareness about measures and methods of reducing GHG emissions and alleviating climate risks in the process of low-carbon economy construction. This is especially true for adaptive activities, which are closely related to people's daily life and property safety. This is a cost effective method of educating about adaptation within low-carbon educational training, which can generate a positive cultural atmosphere and market demand.

(This article was originally published in Chinese in 2012.)

CHAPTER 8

Carbon Dioxide Emissions Policies and Actions of Low-Carbon Development in China's Transport Sector

Cai Bofeng and Feng Xiangzhao

I Introduction

CO_2 emissions in the transportation sector have attracted ever greater attention from both transport and climate change policy experts because of their share of overall emissions and their persistent growth. In 2007, global transportation accounted for 61.2% of the total global oil consumption, equivalent to 2.16 billion tons of standard oil. Transportation has become the largest and fastest-growing oil consuming sector.[1] According to the IEA, 6.62 billion tons of CO_2 were generated by the global transport sector in 2007, accounting for 23% of fossil fuel-related CO_2 emissions.[2] This figure is 1.45 times greater than the 4.57 billion tons emitted in 1990, leading to expectations of reaching 9.30 billion tons by 2030.

In 2008, the US transport sector released 1.795 billion tons (Bt) of CO_2, accounting for 30.32% of total US CO_2 emissions. From 1990 to 2008, CO_2 emissions in the US transport sector rose by 20%.[3] CO_2 emissions in the European Union (EU-15) transport sector reached to 829 Mt in 2008, accounting for 24.98% of total CO_2 emissions.[4] While most of the industrial sectors in EU-15 have successfully abated CO_2 emissions, transport sector emissions increased by 21% in 1990–2008. Transport related CO_2 emissions in developed countries have become an important source of national CO_2 emissions.

* Cai Bofeng, Chinese Academy for Environmental Planning, Ph.D., Email: Caibofeng@gmail.com; Feng Xiangzhao, Policy Research Center for Environment and Economy, Ph.D., Email: feng.xiangzhao@prcee.org.

1 IEA, *CO_2 Emissions from Fuel Combustion 2010*, 2010.

2 IEA, *World Energy Outlook 2010*, 2010.

3 EPA US, *Inventory of US greenhouse gas emissions and sinks: 1990–2008*, 2010.

4 European Environment Agency, *Annual European Union greenhouse gas inventory 1990–2008 and inventory report 2010*, 2010.

 | DOI 10.1163/9789004274648_007

CO_2 emissions in China's transport sector have also dramatically increased because of the rapid growth in motor vehicle ownership. The regional make up and overall estimations of transportation related CO_2 emissions remain uncertain. IEA estimated China's annual transport sector emissions and published the results in its report, *CO_2 Emissions from Fuel Combustion*. The results have been used and cited widely, sometimes by Chinese experts as well. These data remain questionable because no official statistical data or materials directly support the estimates.

According to the *People's Republic of China Initial National Communication on Climate Change*, CO_2 emissions in China's transport sector were 166 Mt in 1994, accounting for 5.40% of total emissions.[5] Since then, very little research concerning China's transport sector CO_2 emissions has been conducted. There was also less research and discussion at the regional and provincial level.

This paper is structured as follows: First, it presents accounts of CO_2 emissions released from China's transport sector, both on the national and regional level. Second, it analyzes the correlation between transport sector CO_2 emissions and socioeconomic factors. Third, it explores the policies which affect low-carbon development in China's transport sector. Finally, it presents policy recommendations for low-carbon development in China's transportation sector.

II CO_2 Emissions in China's Transport Sectors

CO_2 emissions in the transport sector at the national and provincial levels in 2007 were calculated based on provinces' fuel consumption for transportation, transportation turnover, and emissions factors. China's transportation CO_2 emissions in 2007 reached 436 Mt, which is higher than the 408 Mt estimated by the IEA. Figure 8.1 shows the CO_2 emissions share from different modes of transportation. Road transportation accounted for 86.32% of CO_2 emissions, which is much higher than the IEA's estimate of 67.64%.

Water transport CO_2 emissions accounted for 5.49% of total transport sector emissions. Aviation has increased rapidly in recent years, but its overall transport sector emissions share was relatively small, accounting for 5.14%.

5 National Development and Reform Commission, 2004, The People's Republic of China Initial National Communication on Climate Change.

Railway emissions are lower (16.4 Mt) than 2005 levels calculated by He.[6] This is primarily due to some coal-fired steam locomotives that were still in use until 2005 and the quick development of railway electrification after 2005. CO_2 emissions from rail accounted for only 3.05% of total transport emissions, making it the lowest CO_2 emissions contributor.

Provincial CO_2 emissions varied greatly because of different economic development stages and natural conditions. The eastern coastal provinces have higher transport sector CO_2 emissions than those in the western inland regions. However, some eastern provinces, such as Anhui and Jiangxi, also have lower emissions levels, and conversely Xinjiang, a western province, has a relatively high level of CO_2 emission. Guangdong remains the largest transport CO_2 emitter in China, reaching 46.91 Mt. The main reason is that Guangdong is a coastal province and harbors intensive economic activities that leads to a maintained high level of road, water, and air transportation turnover.

Qinghai is among the smallest CO_2 emitting provinces at approximately 1.34 Mt. Tibet's tourism industry is well-developed, and thus its air and road transportation emissions are relatively higher than Qinghai's, although its economic level remains the lowest in China.

CO_2 emissions from air transportation made up a relatively large proportion of total transport emissions in Beijing, Shanghai, and Hainan. CO_2 emissions from water transportation made up a relatively large proportion of total emissions in coastal provinces such as Shanghai, Zhejiang, and Guangdong. The share of rail-related CO_2 emissions was relatively high in some inland provinces such as Hebei, Henan, and Liaoning. CO_2 emissions from road transportation have exceeded 95% of total transport emissions in Tibet, Jilin, Yunnan, and Inner Mongolia.

In Table 8.1, the modal mix of CO_2 emission in the transport sector in China, the international community, and selected countries is compared. Road transportation is overwhelmingly the main contributor of CO_2 emission for the transport sector; it accounted for 94.17% of the European Union's total transport sector CO_2 emissions. The US has an intensive domestic aviation industry, and thus its road transportation share was lower, at about 85.33%. In China, water and rail made up a higher proportion of CO_2 emissions compared with other countries.

---

6 K.B. He, Huo, H., Zhang, Q., et al., 2005, "Oil consumption and CO_2 emissions in China's road transport: current status, future trends, and policy implications," *Energy Policy*, 33 (12): 1499–1507.

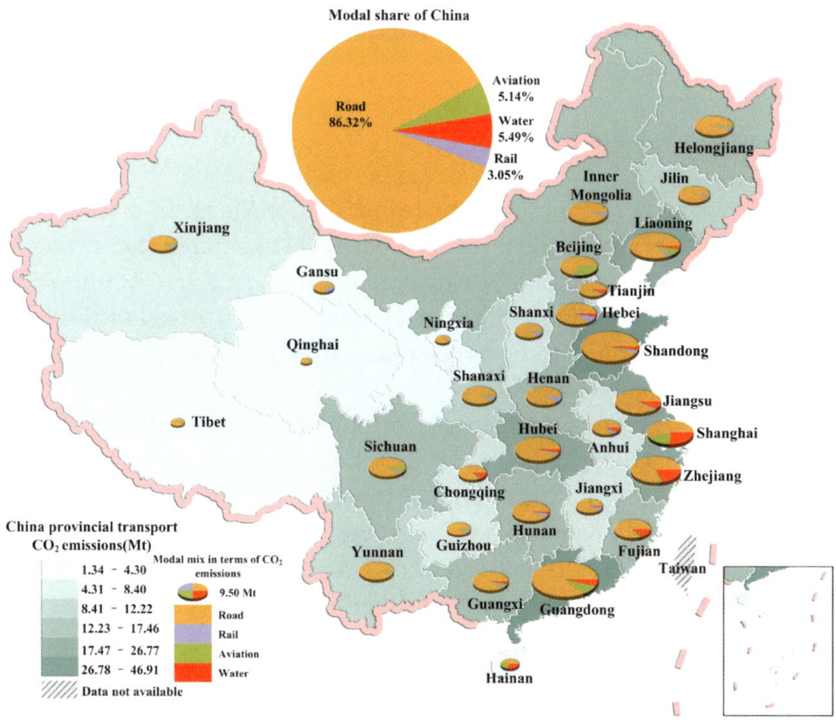

FIGURE 8.1 *Spatial distribution of CO_2 emissions from the transport sector in China by province and modal share*

TABLE 8.1 *Comparison of the modal mix of CO_2 emissions from the transport sector between China and the international community*

CO_2 emissions ratio (%)	Global	Annex I	EU-15	Japan	United States	China
Road	72.81%	88.93%	94.17%	90.04%	85.33%	86.32%
Aviation	27.19%	6.25%	2.62%	4.57%	9.23%	5.14%
Water		2.77%	2.54%	5.12%	2.93%	5.49%
Rail		2.05%	0.67%	0.27%	2.51%	3.05%

Note: Annex I refers to the Annex I countries under the Kyoto Protocol.

III Potential Driving Forces behind China's Transport Sector CO_2 Emissions

Timilsina (2009) indicated that GDP and per capita GDP are the most important factors influencing CO_2 emissions in the transport sector.[7] Using transport sector CO_2 emissions from different region and provinces in place of time series data allows us to analyze the relationship between transport CO_2 emissions and GDP and per capita GDP on a spatial level. As Figure 8.2 shows, the relationship between transportation CO_2 emissions and GDP is statistically significant ($R^2 = 0.811$) and stronger than that between transportation CO_2 emissions and per capita GDP, which is virtually insignificant ($R^2 = 0.214$). This is because the main forces driving transportation CO_2 emissions are industrial and economic activity rather than household incomes. China's economic development depends on industrial production, export, and infrastructure construction on a fundamental level and does not depend in that way on residential consumption. The per capita GDP in Guangdong is lower than in Beijing, Shanghai, Zhejiang, and some other provinces, but its CO_2 emissions from the transport sector are the highest nationwide. The per capita GDP in Tianjin is higher than China's national average, but its transport sector CO_2 emissions are relatively low. The strong correlation between transportation CO_2 emissions and regional GDP generally demonstrates that CO_2 emissions from the transport sector in China are predominantly driven by the intensity of production activities rather than consumption activities.

Road transportation CO_2 emissions are generally understood to be significantly affected by residents' income because private cars comprise the majority of vehicles and car ownership increases with income. However, according to the statistical analysis at the provincial level, CO_2 emissions from road transportation were only loosely associated with residents' income ($R^2 = 0.147$) but closely associated with GDP ($R^2 = 0.794$, Figure 8.3). Thus, private cars may not be the sole primary CO_2 emissions source. Trucks, taxis, company owned vehicles, and government owned vehicles may contribute significantly to total road emissions. Analysis of transport sector and road transportation CO_2 emissions crossed with population, in Figure 8.4, indicates that there is no strong correlation, proving that CO_2 emission in China's transport sector are deeply connected to the intensity of economic activities rather than to population.

7 Timilsina G.R., Shrestha A., "Transport sector CO_2 emissions growth in Asia: Underlying factors and policy options," *Energy Policy*, 2009, 37 (11): 4523–4539.

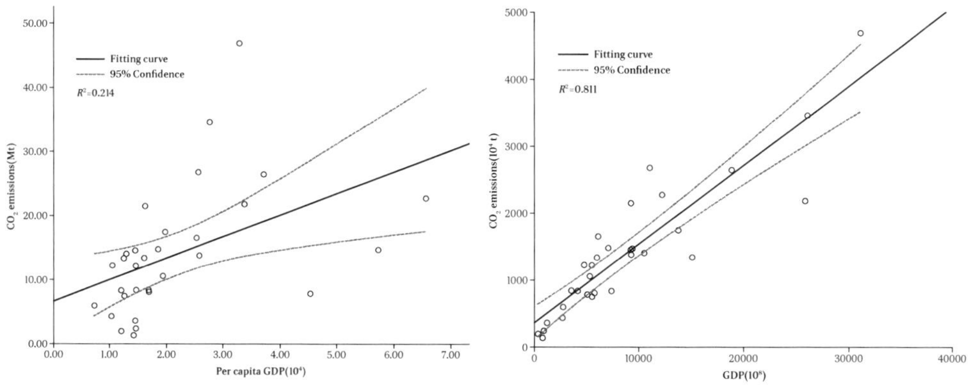

FIGURE 8.2 *Correlation between transportation CO_2 emissions and per capita GDP/GDP*

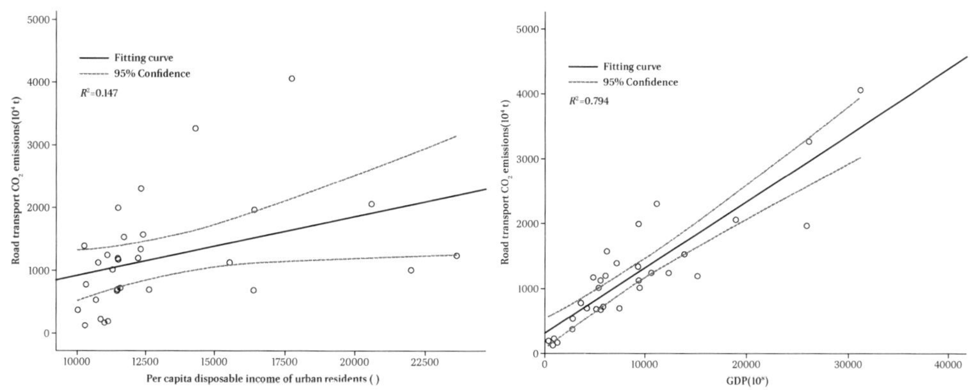

FIGURE 8.3 *Correlation between road transportation CO_2 emissions and per capita disposable income of urban residents and GDP*

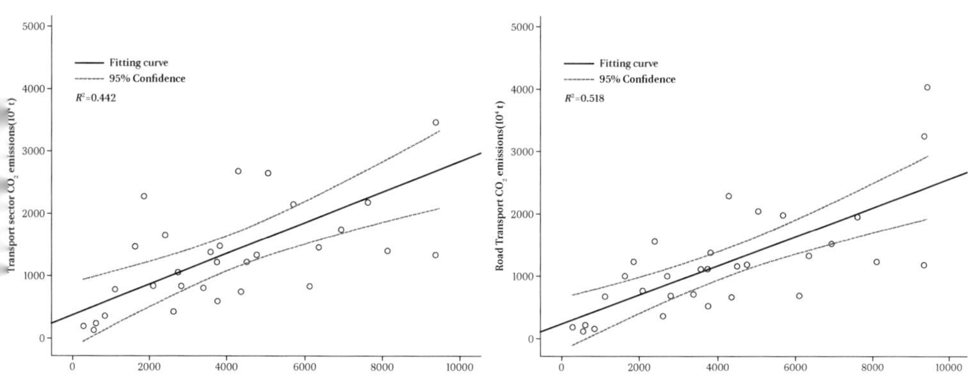

FIGURE 8.4 *Correlation between transport/road transport CO_2 emissions and population*

IV Low-Carbon Policies and Actions in China's Transportation Sector

Experiences in many countries have proven that the most significant driving factor in facilitating carbon emissions reduction and furthering low-carbon development in the transportation sector is innovation and development of systems and policies. China has yet to issue a systematic low-carbon transportation strategy and few policies directly address CO_2 reduction in the transportation sector. Low-carbon transportation policies, defined narrowly, are lagging. However, China has formulated various policies in order to alleviate urban traffic congestion, improve air quality, guarantee transportation energy supply, and facilitate sustainable transportation development. Those policies have played an important role in reducing China's transportation sector carbon emissions. Therefore, using a broad conception of low-carbon transportation policies and practices, China had many examples to show.

Due to the dominant role of road-related CO_2 emission in the transportation sector at home and abroad, an overview and summary of China's low-carbon transportation policies gives priority to road transportation, and looks at railway, civil aviation, and water transportation as a supplement.

A *Road Transportation*

The road transport sector is considered to be a complex system that is comprised of four basic elements, namely people, vehicles, fuels, and roads. All the current road transport sector policies and measures consider those basic elements to be the main control objectives. Based on observed patterns, policies and measures generally fall into the following three categories. First, the government issues command and control policy instruments that generally comprise of laws, regulations, standards, and planning such as standard limits on passenger vehicle fuel consumption and the development planning in the "10th Five-Year Plan" on denatured ethanol fuel for motor vehicles. Second, the government uses economic incentives including fuel taxes and fiscal subsidies for R&D on new energy technology for vehicles in the National 863 Research Project. Third, the government uses persuasion and education including "Green Travel Week" and "Car-free Day" that have been promoted since 2007 in China.[8] The following table shows some low-carbon policies and practices in China's transportation sector policy matrix. The main low-carbon policies and measures that are currently being implemented will be discussed further.

---

8 Feng Xiangzhao, *A Strategic Study on How to Reduce GHG Emissions from the Urban Transport Sector* (Beijing: China Meteorological Press, 2009).

TABLE 8.2 *Low-Carbon Policy Matrix in China's Road Transport Sector*

Policy type/ transportation element	Vehicles	Fuels	Roads	People
Command and Control	Fuel economy standard (Passenger car fuel consumption limit standard, Commercial vehicle fuel consumption limit standard); systems of auction and lottery on license plate of motor vehicles; measures restricting transportation of motor vehicles; *Administrative Measures on the Inspection and Supervision of Road Transportation Vehicles Fuel Consumption; Limits and Method for Measuring Commercial Vehicles' Fuel Consumption*	Development planning in the "10th Five-Year Plan" of denatured ethanol fuel for motor vehicles; *GB18350-2001* standards for denatured ethanol fuel; *GB18351-2001* standards for ethanol gasoline for motor vehicles	BRT; immediate processing methods for road transportation accidents	Flexible working systems (system of staggered working hours)
Economic incentives	High downtown parking fee; tax breaks for energy efficient and environmentally friendly vehicles; destruction of yellow label vehicles and policies providing fiscal subsidies for auto replacement; vehicle and vessel tax	Fiscal subsidies for R&D (support for R&D on alternative fuels, hybrid vehicles and electric vehicles through the National 863 Project); fiscal subsidies for manufacturing and alternative fuel use (Notice on the subsidy policy for ethanol fuel); fuel tax; tax breaks		Corporate commuting subsidy; low-ticket-price system for public transit and subways; parking fee reduction at Park & Ride

TABLE 8.2 *(cont.)*

Policy type/ transportation element	Vehicles	Fuels	Roads	People
Persuasion and education	*Labels indicating Fuel Consumption for Light Vehicles; Mark of Conformity for Motor Vehicles with Environmental Protection*	Air purification projects and actions for sanitary cars		Green Travel Week; Car-free Day

1 Fuel Economy Standards

Fuel economy standards are considered to be one of the most effective instruments for controlling fuel consumption and carbon emissions at home and abroad. Figure 8.5 shows both actual fuel efficiency data in different countries and regions and the fuel economy standard targets from plans yet to be implemented. The figure shows that the European Union and Japan have the highest fuel efficiency for passenger vehicles and that those two have the lowest CO_2 emissions for single motor vehicles.

Since 2005, China has implemented the *Passenger Car Fuel Consumption Limit Standard GB19578-2004* that places a minimum required fuel economy for new passenger vehicles. *GB19578-2004* was implemented in two phases: the first phase began on July 1, 2005 at which point all new vehicles were required to meet the standard and all old vehicles had one year to come into compliance; the second phase began on January 1, 2008 once again placing the requirement immediately upon all new vehicles and giving a one year leeway for old vehicles. On July 19, 2007, China issued the *Light-Duty Commercial Vehicle Fuel Consumption Limit Standard,* which was implemented on February 1, 2008. The implementation of these standards has saved 910,300 tons of gasoline and 680,600 tons of diesel, which is equivalent to a reduction of 5,149,400 tons of CO_2.

China's current fuel economy standards also include testing methods, systems for examination, approval, and publicity, label management, supervisory mechanisms, as well as assorted administrative measures concerning rewards and punishment.[9] *Fuel Consumption Labeling for Light-Duty Vehicles* was

9 Feng Xiangzhao, Zou Ji, Xu Guangqing, "An Economic Review on Current Fuel Economy Standards in China," *Environmental Protection*, 2008, (6), 23–26.

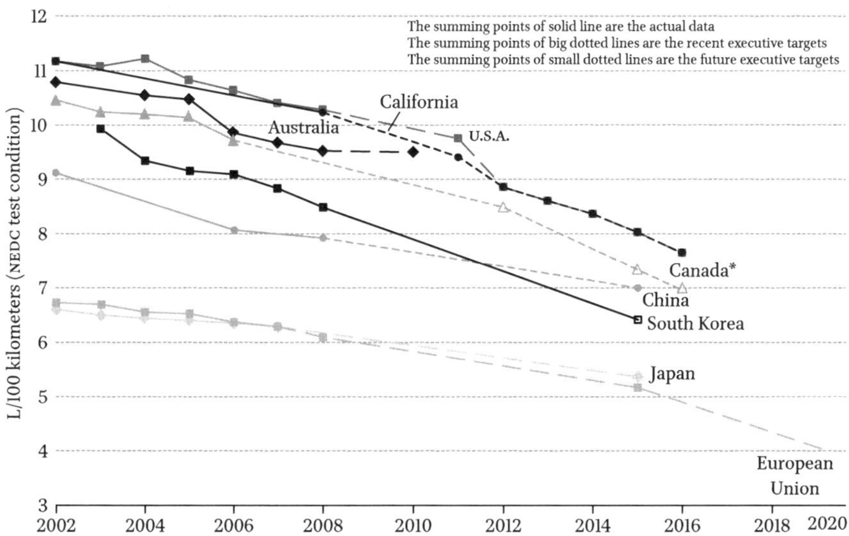

FIGURE 8.5 *Comparison of Passenger Vehicle Fuel Economy Standard Targets in Different Countries*
Notes: * The target value for Canada for 2016 is the estimated value from the ICCT rather than the official figure

implemented in China on January 1, 2010 to help guide the public in selecting energy efficient and environmentally friendly vehicles and to improve the public's low-carbon environmental awareness.[10]

Figure 8.6 provides a comparison between China's CO_2 emission standards that are derived from China's fuel economy targets and international standards. According to this figure, developed countries have great power to control motor vehicles' CO_2 emission and the US has the largest range of decrease. Compared with developed countries, China's target in 2015 still has a long way to go.

According to the latest research from the International Council on Clean Transportation (ICCT), if China can achieve notable reductions in CO_2 emissions if it applies the improved scenario (Euro IV is implemented in 2015 and the standard on light vehicles is further restricted by 3% each year) and the strong scenario (Euro V in 2012, Euro VI in 2015, "ultra-low emissions" for light-duty vehicles, and Euro VII for heavy-duty vehicles are put in place in 2020) to control the emissions of motor vehicles (Figure 8.7).

10 Wang Zuohan. "Fuel Consumption Labels for Light Vehicles. Compulsory Execution in 2010," *Commercial vehicles*, 2009, 9.

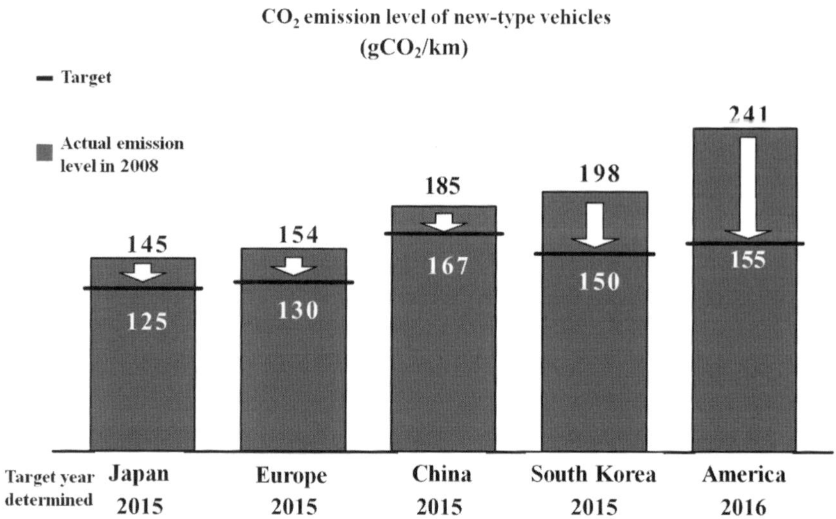

FIGURE 8.6 *Current status and targets of CO_2 emission from new light passenger vehicles in major countries*[11]
Note: Data for China is derived from the new version of the *Passenger Vehicles Fuel Consumption Limit Standard* that is expected to be published soon and implemented in 2015. In the new version, the average fuel economy of passenger vehicles in China will decrease to around 7L/100km and CO_2 emissions decrease to 167g/km.

The *Development Plan of the "12th Five-Year Plan" for Transportation,* which was issued in the first half of 2011 in China, specifies that during the 12th Five-Year Plan the unit energy consumption of transportation turnover for passenger vehicles and freight vehicles will be reduced by 6% and 12% respectively and the energy consumption of commercial vehicles and CO_2 emissions will decline by 10% and 11% respectively compared with 2005 levels.

2 Policies Promoting the Development of Alternative Fuels

Alternative fuels are considered by all countries as an important strategy for reducing CO_2 emission in the transportation sector. China has established preliminary policy systems to encourage the development of alternative fuels and vehicles to consume them guided by national macro-policy and supported by specific local policies.

11 Partial reference for the Climate Group. Low-carbon Technical Marketing for Electric Bicycles, 2010.

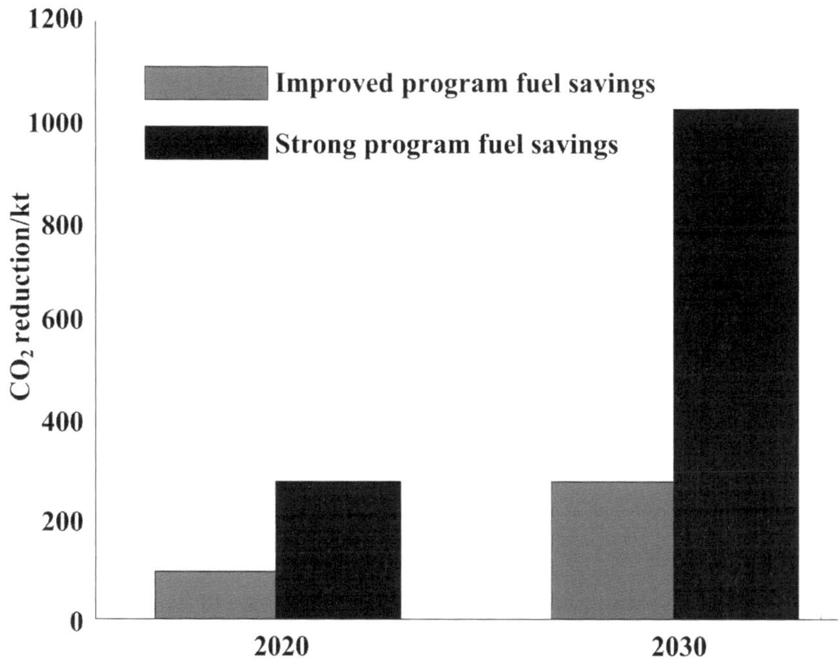

FIGURE 8.7 *CO_2 emissions mitigation under different fuel economy scenarios in China*

Besides the *Energy Conservation Law* and the *Renewable Energy Law* in which the strategic plan for developing alternative vehicle fuels is laid out, alternative fuels' significance to low-carbon transportation development has been emphasized many times in development plans that were consequently issued by the Chinese government since the new century, such as the *National Energy Medium-and Long-term Development Plan for 2004–2020*. In order to accelerate the application and dissemination of ethanol fuel, China has issued two important standards in 2001, namely, the *Denatured Ethanol Fuel GB18350-2001* and the *Ethanol Gasoline for Motor Vehicles GB18351-2001*.[12]

Policies such as fiscal subsidies and various tax breaks provide further important economic incentives to promote the development of alternative fuels. In June 2004, the Ministry of Finance issued a *Notice on the Subsidy Policy for Fuel Ethanol*, which implements policies of finance, taxation, and pricing on preferential subsidies for the production and utilization of ethanol fuel. A 5% consumption tax exemption has been made for the four assigned

12 Feng Xiangzhao, *Strategic Research on GHG Emissions Reduction in the Urban Transportation System*, (Beijing: China Meteorological Press, 2009).

ethanol fuel manufacturing enterprises. The value-added tax on ethanol fuel manufacturing reimbursement scheme has already been implemented. Subsidies for aged grain for ethanol fuel production and price discounts on stale grain supply have also been put in place. Some local governments have given preferential business income tax rates for ethanol manufacturing enterprises and refueling stations.

3 Economic Incentive for New Energy Vehicles

Under pressure to save energy and reduce carbon emissions, the development of new energy vehicles has become a significant opportunity for the global automotive industry. In order to stimulate the development support for key technology, R&D on new energy vehicles must be strengthened and economic incentives that push through the commercialization and industrialization of these technologies must be enacted.[13]

Since 2005, China has pinpointed ways to facilitate the development of the new energy vehicle industry in many plans, and issued a series of preferential policies meant to accelerate the industrialization of new energy vehicles. On January 23, 2009, the Ministry of Finance and the Ministry of Science and Technology jointly issued the *Notice on Launching Pilot Work Regarding Demonstration and Dissemination for Energy-Saving Vehicles and New Energy Vehicles*, and initiated the project called "Thousand Vehicles in Ten Cities." The project launches demonstration and dissemination pilot projects of energy-saving vehicles and new energy vehicles in 13 selected cities including Beijing, Shanghai, and Chongqing, and encourages them to lead in energy-saving and new energy vehicles use in public service sectors such as public transportation, taxis, government departments, environmental sanitation, and the postal service. This would be achieved via fiscal policies, provision of subsidies for organizations that purchase energy-saving and new energy vehicles, and an attempt to enlarge the operating scale of China's new energy vehicles to 10% of the automobile market share by 2012. In May 2010, another seven pilot cities were added including Tianjin and Haikou. In total, 20 Chinese pilot cities have been involved in the low-carbon transportation initiative.

In June 2010, China officially initiated pilot projects testing subsidies of new energy vehicles for private purchase by providing direct subsidies. It confirmed five cities to launch the new energy vehicles for private purchase subsidies pilot projects, covering Shanghai, Changchun, Shenzhen, Hangzhou, and Hefei (Beijing was later added). The subsidy standard is based on the energy of the

13 Jia Xinguang, "Strengthening to Implement the Subsidy Policies for New Energy Vehicles," *Economy*, 2010, (8): 110.

battery pack power, in which 3000 RMB/kWh will be subsidized for those new energy vehicles that satisfy the required conditions.

4 Preferential Policies for Fuel Taxes and Vehicle Taxes

Fuel taxes are also regarded as important policy tools to effectively guide consumers to appropriate use of vehicle fuels, reduce energy consumption from road transportation, and lower carbon emission. In China, fuel taxes refer to the taxes that are levied on gasoline and diesel used for vehicles. The fuel tax introduced on January 1, 2009 nullified the other six fees which were levied beyond the price of oil products, which included road tolls, waterway maintenance costs, administrative fees for road transportation, additional charges for passenger and freight transportation on roads, administrative fees for water transportation, and extra charges for passenger and freight transportation on waterways. The tax per unit of gasoline consumed, levied based on the price of oil products, has increased by 0.8 RMB/unit, from 0.2 RMB/unit to 1 RMB/unit, and that of diesel has increased by 0.7 RMB/unit, from 0.1 RMB/unit to 0.8 RMB/unit.

Excise taxes, vehicle purchase taxes, as well as vehicle and vessel taxes are also important means for policy makers to facilitate energy savings and emissions reductions in the transportation sector. Since 2006, China has consistently raised the excise tax rate on products such as vehicles with large engines, and implemented tax breaks for vehicles with small engines. On January 20 2009, the Ministry of Finance and the State Administration of Taxation jointly issued the *Notice on Implementing the Reduction of Vehicle Purchase Tax Policy*, proposing that if passenger vehicles smaller than 1.6 L (including 1.6 L) were purchased from January 20 to December 31, 2009, the vehicle purchase tax would be reduced by 5%. In February 2011, China issued the *Law on Vehicle and Vessel Taxes,* which was implemented on January 1, 2012. The vehicle and vessel tax is based on a progressive increase of engine size, and is very similar to the Vehicle Carbon Tax that is implemented in European countries. For the purposes of this article, the vehicle and vessel tax will be expressed as the equivalent level of carbon tax, which means that if the average CO_2 emissions per vehicle in China is 160g/km, the vehicle and vessel tax levied is equivalent to the carbon tax level of 11 RMB for 1 ton of CO_2.

5 Other Low-Carbon Policies and Actions

i. *Integration of Transportation Planning and Urban Planning*

Dynamic integration of urban master planning and transportation planning across multiple levels is considered to be an important safeguard for effectively alleviating urban traffic congestion and improving the urban transportation system. The dynamic combination of transportation planning and urban planning is championed by TOD (transit-oriented development guided by the public transportation). TOD, an urban development mode guided by public transportation, is the brain-child of American urban planning, which seeks to make high-density and multifunctional development for large-capacity public transit routes by integrating the planning of urban land and urban transportation so as to decrease people's daily travel demand, establish large-capacity travel by means of public transportation, effectively increase the operating efficiency of the transportation system, reduce transportation based energy consumption, and lower the transportation system's carbon emissions.

Beijing has begun exploring the integration of transportation planning and urban planning. For example, in accordance with the *Beijing Urban Master Plan* (*2004–2020*) and the *Outline of Development for Beijing Transportation*, Beijing has launched a compilation of special plans such as road transportation plans, light rail transit plans, highway network plans, comprehensive transportation plans for public transportation networks, and operating organizations.

ii *Rail Transit and BRT*

In order to relieve urban traffic congestion and facilitate the sustainable development of urban transportation, the construction of rail transit has entered a new stage of rapid development in recent years in large and medium cities. By 2008, the operating length of urban rail transit was 775.6 kilometers nationally. Over 1800 kilometers are under construction, involving 15 cities with construction already begun, 10 cities still waiting for approval, and seven cities preparing for construction.

Bus Rapid Transit (BRT), which is a large-volume rapid system that offers convenience, reliability, comfort, and low cost, has also been rapidly developing in China recently. From December 30, 2005 when the 1st Beijing BRT line began operation, through to the end of 2008, China has built 20 BRT lanes in 10 cities. Beijing, Jinan, and Xiamen have three lanes each, Hangzhou, Hefei, Kunming, and Changzhou have two lanes each, and Dalian, Zhengzhou, and Chongqing have a single lane each.

iii *Policies Limiting the Purchase and Use of Motor Vehicles*

Limiting the purchase and use of motor vehicles is another low-carbon policy choice adopted by many urban administrative departments in China and abroad, which seeks to alleviate urban traffic congestion and improve air quality in urban areas. In China, most limitation policies use motor vehicle license plate auction and lottery systems and a transportation restriction system.

Shanghai is the only city in China where a license plate auction system for private cars has been implemented. China began mass control over the quota of newly added private cars in downtowns through bidding and auction. Because of the implementation of the license plate auction system, the number of private motor vehicles in Shanghai remains low compared with Beijing.

Using a different strategy, Beijing implemented a motor vehicle quota on January 1, 2011, which is also known as a motor vehicle license plate lottery system. In accordance with the *Interim Provisions on Passenger Vehicles in Beijing Quality Control* and its enforcement regulations, the annual quota for passenger cars in Beijing is set at 240,000 for 2011. These will be distributed for free through a license plate lottery that seeks to be open, fair, and just. On July 11, 2011, the Guiyang Municipal Government published its *Interim Provisions on Passenger Vehicle License Plate Administration in Guiyang City*, proposing that the number of new licensed passenger cars (passenger cars with less than 9 seats,) which travel within the 1st ring road (including the 1st ring road,) should be restricted starting July 12, and that 3000 additional passenger cars should granted access every month via the license plate lottery.

Aside from measures that limit vehicle purchase, some cities have implemented measures that place driving restrictions based on the last number of the license plate. In Beijing as example, these measures have been in operation since 2008, when an odd-and-even license plate rule was passed during the Olympic Games, and have been updated to the current one car-free day per week.

iv *Promotional and Educational Activities Relating to Energy Saving and Environmental Protection in the Transportation Sector*

Promotion and education as well as information dissemination is another standard policy choice for governments seeking to change people's improper consumption patterns. Because public transportation and cycling benefit from low cost and low emissions (as shown in Figure 8.8), it is also the important for the development of low-carbon transportation to encourage people to adopt green travel options via bicycles and public transportation.

In November 2006, the former Ministry of Construction advocated for the launch of "China's urban public transportation week and car-free day" throughout the nation, and hoped to accelerate the construction and development of urban public transportation and sustainable urban transportation systems through the joint participation of governments, enterprises, media, and the public. 110 cities nationwide responded to the proposal and signed the *Letter of Commitment to Launching China's Urban Public Transportation Week and Car-Free Day*. The first campaign was launched from September 16–22, 2007,

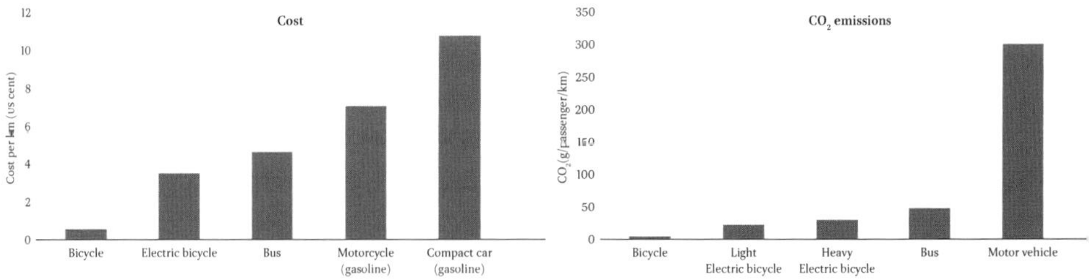

FIGURE 8.8 *Travel Costs and CO_2 Emissions for Different Transportation Modes in China*[14]

adopting the theme of "Green Transportation and Health." Those campaigns have propelled energy savings and emissions reductions in the transportation sector and improved urban air quality. Within the 1st ring road in Kunming, on car-free day, the exhaust emissions of CH, NO_2, CO, CO_2 as well as PM were reduced by 55.9%, 45.1%, 32.9%, 15.4%, and 20.7% respectively compared with the previous day.[15]

B *Low-Carbon Development of Other Transportation Modes*

1 Railway Transportation

Railway transportation has ushered in unprecedented development during 2006–2010 in China, allowing for advances in technology that leapfrogged struggles faced by other nations. By 2010, the length of railway in operation in China increased by 21.3% from the end of 10th Five-Year Plan period to 91,000 kilometers, ranking the second longest in the world. High-speed railway construction has also achieved some outstanding feats. The length of high-speed railway in operation has reached 8,358 kilometers in China, the longest in the world. 1,200 motor trains are driven every day and large tracts of high-speed railways are under construction.[16]

Through the structural reform of locomotive traction power, the priority of steam locomotive has shifted to an emphasis on both diesel and electric locomotives. The railway sector's energy consumption structure has been advanced

14 The World Bank, *World Development Report: 2010 Development and Climate Change*, 2009.

15 Jiang Yulin, Wu Hongxiang, Shen Yun, *Smooth, High-efficient, Safe, and Green-Analysis of Major Issues of Sustainable Development for China's Urban Public Transportation*, (Beijing: Science Press, 2010).

16 "Comprehensive Views: Investment in Chinese Railway in 2011 is not Less Than That in Last Year," Hexun Website, Feb. 24, 2011. See details at: http://futures.hexun.com/2011-02-24/127540850.html.

from using predominantly coal to electricity and oil. During the 11th Five-Year Plan period, the production of new railway lines reached 14,700 kilometers, double that of 10th Five-Year Plan, and the production of multiple tracks and electrification has reached 11,200 kilometers and 21,300 kilometers respectively, which is an increase by a factor 3.1 and 3.9 respectively compared with the size during the 10th Five-Year Plan. System-wide multiple track rate and electrification rate has reached 41% and 46% respectively.[17]

By improving diesel locomotives' technical equipment and developing electronic fuel injection technology, the fuel economy level of diesel locomotives has been greatly enhanced. By electrifying existing railway, the traction share of electric locomotives has increased. In the course of the 11th Five-Year Plan, the energy consumption level per unit freight turnover of railway has experienced declining momentum, and the comprehensive energy consumption of locomotive unit freight capacity has decreased to 4.9 tons of standard coal equivalent per million ton-kilometer, from the 5.73 tons of standard coal equivalent per million ton-kilometer seen in 2005.

2 Air Transportation

In order to deal with the pressure to reduce emissions in China and abroad and reduce the increasing burden of aviation fuel cost, China's aviation industry has taken various effective measures to reduce aviation fuel consumption and lower the expenditure on aviation fuel. Using current aircraft technology, fuel savings across the entire operation have been realized by employing scientific flight patterns, operation potential maximization, and security maintenance. For example, fuel saving from production and operation can be realized by better matching machine types, air lines, aviation markets, and using machines with a comparatively low unit fuel consumption to carry out air line's prerequisites and finishing the voyage and payload. Energy savings and emissions reductions for airlines demand the full cooperation across departments such as air traffic control and airports. Civil aviation planes have been granted straight channels by opening previously restricted air paths and exploring new routes so as to reduce flight time and decrease aviation fuel consumption. Improvements in navigation management allow for a well-organized and smooth air traffic environment that reduces unnecessary circling in the air and queue time on ground. In the airport, proper designs of departure lounges, gate positions and runways will shorten queue time of both inbound and

17 Ibid.

outbound planes. These measures have laid a solid foundation for realizing energy savings and emissions reductions in the aviation sector.[18]

The European Commission issued a new act (2008/101/EC) on November 19, 2008, which brings international aviation into the European Emissions Trading Scheme (EU ETS). Starting January 1, 2012, all flights that fly to or pass through European Union airports will be party to EU-ETS, and the emissions of airlines for 2012 will be restricted to 97% of the average emissions from 2004 to 2006. These regulations have placed new challenges for China's low-carbon development of air transportation.

During the11th Five-Year Plan period, the ton-kilometer fuel consumption (kilogram) of CAAC has decreased from 0.336 kg in 2005 to 0.306 kg, which is equivalent to CO_2 emissions reductions of 4.97 million tons compared with a baseline scenario for 2010. In accordance with the *12th Five-Year Plan for the Development of China's Civil Aviation*, during the 12th Five-Year Plan period, the average ton-kilometer fuel consumption for civil aviation will decline from the 0.306 kg in the 11th Five-Year Plan period to below 0.294 kg.

3 Water Transportation

China's water transportation sector has seen regulations such as the *Methods on Implementing the 'Energy Saving Law of the People's Republic of China in the Fields of Road Transportation and Water Transportation* over the past few years. Furthermore, Standards like *Design Specifications for Energy Savings in Marine Traffic Engineering* have been formulated; *Outline of Long and Medium Term Energy Savings in the Fields of Road Transportation and Water Transportation* and other key points on annual energy saving and emissions reduction have been published; guidance documents such as *Guiding Opinions for the Transportation Industry on Comprehensively Implementing the 'Decision on Strengthening Energy Saving Work by the State Council'* and *Development Policies for Resource Efficient and Environmentally-Friendly Road Transportation and Water Transportation* have been printed and distributed.

In addition, work on saving energy and reducing consumption and emissions has also progressed by strengthening water transportation trade management, establishing energy saving and emissions reducing supervisory systems, enhancing the concentration of shipping enterprises, accelerating marine technology upgrades, promoting conversion from fuel to electricity for gantry cranes of rubber-typed containers in harbors (RTG), and exploring the application of shore power technology for anchored ships. According to the *12th Five-*

18 Chai Yufeng, "Thinking of the Work on Energy Saving and Emission Reduction for China's Airlines under the New Situation," *Air Transport & Business*, No. 16 (2009), 4–7.

Year Plan for Energy Saving and Emissions Reduction in the Fields of Road Transportation and Water Transportation, during the 12th Five-Year Plan period, the unit energy consumption of transportation turnover for ships in operation will decrease by 15% and CO_2 emissions will decrease by 16% compared with 2005 levels, and the unit energy consumption of transportation turnover for ships in operation at sea and in inland rivers will decline 16% and 14% respectively.

V Policy Recommendations and Prospects for China's Transportation Sector's Low-Carbon Development

Although some policies and actions promoting low-carbon transportation have already been undertaken in China, carbon emissions in the transportation sector maintains rapid increasing momentum, leading it to be one of the energy use categories with the fastest increasing carbon emissions. Carbon emissions reduction in this sector must be taken seriously, and China should work to build and refine low-carbon transportation policies and strive to save energy, reduce consumption and emissions, and facilitate sustainable transport development. The following proposals outline steps forward with regard to road transportation.

First, China should evaluate the effect of implementing immediate limit standards on fuel consumption for passenger vehicles and future standards for light commercial vehicles, strictly implement the standards deemed appropriate and update them as needed, further on when the system has adjusted, China should apply limit standards on fuel consumption for heavy commercial vehicles as well, and in doing so, facilitate the overall enhancement of motor vehicles fuel economy. A further step is to bolster projects that eliminate obsolete vehicles, reinforce the implementation of preferential measures that support purchase of energy-saving and environmentally-friendly vehicles and of fiscal preference policies for auto replacement, which will aid in realizing energy savings and reduction of NO_x emissions during the 12th Five-Year Plan period.

Second, to address alternative fuels and new energy vehicles, China should formulate medium and long-term development plans, make use of industry associations, organize technology innovation alliances in the electric vehicle industry, accelerate the formulation of relevant standards and access policies, further support R&D and industrialization of alternative fuels and new energy vehicles, and issue medium and long-term infrastructure plans for alternative fuel vehicles and electric vehicles.

Third, the low-carbon transport development can be pushed forward by taking advantage of tax leverage and fiscal instruments. Reforming oil products' pricing mechanism can be used as an opportunity to refine the fuel tax system and raise funds for urban road transportation system construction. With environmental tax reform, China can conjure an adjustment toward low-carbon activity using purchase taxes and consumption taxes levied on motor vehicles, and it can further conduct active research of the feasibility of a carbon tax on motor vehicles.

Four, the status of transportation planning is to be further enhanced by incorporating it within urban planning. China should strengthen the harmonious and cooperative development of urban land-use and urban transportation, strive to refine pertinent laws and regulations, and actively promote the practice of TOD.

Fifth, the construction of urban rail transit infrastructure is to accelerated. China should increase the supply of public transit facilities, improve the operating efficiency of urban road transportation systems, alleviate traffic congestion, facilitate transportation energy savings, and reduce the systematic level of carbon emissions.

Finally, in order to further improve the public's awareness of low-carbon environmental projects, and advocate for green travel options, meaning production of a small carbon footprint to satisfy travel demands, active promotion and education efforts should be undertaken.

In the railway, aviation, and water transportation sectors, China can build onto and refine the comprehensive low-carbon transportation system, promote realization of energy saving and carbon reduction targets set for the 12th Five-Year Plan period, and seek positive strategies for coping with the serious challenges brought forth by climate change while satisfying the increased transport demand through bolstered management of energy saving and carbon emissions reduction requirements, optimization of the energy structure, implementation of major energy saving and emissions reduction projects, energy utilization efficiency enhancements, active support of basic research of key energy saving and emissions reducing technologies, and acceleration of their distribution and commercialization.

(This article was originally published in Chinese in 2012.)

CHAPTER 9

Demand for and Policies to Improve Low-Carbon Financing

Pan Jiahua, Hongbo Chen, Xiang Yu and Lijuan Wang

Abstract

This paper conducts an analysis of the development of major financing facilities such as public fiscal funds, financial institutions, the capital market, and the carbon market in China, and assesses the effect of implementing existing policies to support carbon finance. It presents policy-related advice on how to achieve targets on the basis of a forecast of low-carbon development's costs and capital demands in China up to 2020. According to the research, costs of emissions reduction are growing year by year in China and will reach 2.47% of GDP by 2020. Corresponding capital demand will total RMB 6.8–18.12 trillion where RMB 0.91 trillion will be needed annually from 2011 to 2015 and RMB 1.36–3.6 trillion will be needed annually from 2016 to 2020. Such a tremendous capital demand requires not only active public financing but also a multi-channel, multi-level and multi-faceted financing system.

Keywords

low-carbon financing – public financing – carbon market – capital demand – emissions trading

* Pan Jiahua is general director at the Institute for Urban and Environmental Studies, the Chinese Academy of Social Sciences (CASS), research fellow and doctoral supervisor, specializing in world economics, climate change economics, urban development, energy and environmental policies, etc. Hongbo Chen is an associate research fellow at the Institute for Urban and Environmental Studies, CASS, specializes in environmental economics, carbon market and climate change policies, energy conservation for buildings, etc. Xiang Yu is a doctor at the Institute for Urban and Environmental Studies, CASS, where he specializes in carbon financing and climate change policies, corresponding author of this paper. Lijuan Wang is a doctor at the Institute for Urban and Environmental Studies, CASS and specializes in energy economics and climate change policies.

 | DOI 10.1163/9789004274648_010

China faces challenges even greater than those faced by developed countries in mitigating climate change due to its national conditions and the nature of its current evolutionary stage. China will maintain its focus on industrialization and urbanization during the upcoming decade. It will therefore be crucial for it to increase energy efficiency, develop new and renewable energies, reduce pollution discharge, and improve the quality of the environment in order to lay the foundation for China's economic transformation and to begin the gradual realization of low-carbon growth. The Chinese government announced in 2009 that it aims to reduce its carbon dioxide emissions per unit of gross domestic product (GDP) by 40–45% from 2005 levels and increase the share of non-fossil energy in its primary energy consumption up to around 15% by 2020. China's 12th Five-Year Plan further sets targets to reduce energy intensity by 16%, to cut down carbon intensity by 17%, and to increase the share of non-fossil energy in primary energy consumption to 11.4% by 2015. Realizing these targets requires large-scale technical reform of existing industries and tremendous investment in new and renewable energy industries, which implies enormous fund raising and input efforts.

This paper forecasts the required amount of financing, analyzes the status of major financing channels such as public fiscal funds, financial institutions, the capital market, and the carbon market in China, and evaluates the effect of implementing existing policies addressing the low-carbon development targets already declared the Chinese government, offering advice on China's long-run low-carbon financing policies.

I Challenges and Difficulties for Low-Carbon Financing

Low-carbon financing refers to fund raising activities aimed at combating climate change, promoting the decarbonization of the economy, and boosting sustainable development. Specific examples include: (1) developing new and renewable energies such as wind energy, solar energy, biomass energy, nuclear power, and hydropower; (2) conserving energy in industries with traditionally high energy consumption such as power, steel, petrochemical, chemical, construction materials, machinery, light industry, and non-ferrous metallurgy; and (3) conserving energy used by buildings, transportation, and the daily lives of Chinese people.

Low-carbon financing channels in China come from one of two sources, international and domestic. International climate change financing is critical to global actions against climate change. Its key objectives are to facilitate climate adaptability improvements in developing countries, cut down greenhouse gas (GHG) emissions, and achieve sustainable development. Capital for the purpose of international climate financing is sourced predominantly from funding mechanisms developed by the United Nations Framework Convention on Climate Change (UNFCCC), such as the Clean Development Mechanism (CDM) under the Kyoto Protocol, and other bilateral and multilateral funds. China's domestic low-carbon finance includes all locally-sourced capital that aims to promote a sustainable model of economic growth.

Domestic finance is mostly comprised of public financing from the government, financing from financial institutions, equity financing from the capital market, debt financing, and financing from the carbon market. The money sourced through international climate financing has so far been insufficient to stimulate the necessary climate change mitigation and alleviation practices in developing countries, including China. A variety of bilateral and multilateral funding mechanisms exist outside of the UNFCCC and are decentralized and not bound by the UNFCCC. Donations to these funds depend on donors' willingness and ability to provide, hence financing is not guaranteed. With the European debt crisis, the slow recovery of the US economy, and the continuing gloom of the Japanese economy, developed countries are showing less and less political will to sponsor climate change projects in developing countries. Although China occupies the largest share in the CDM market and has been proactively seeking opportunities in the international carbon market, the volume of finance is also much smaller than the amount needed by China and other developing countries due to the low and sharply fluctuating price of certified emissions reductions (CERs).

GHG emissions across the globe were historically largely generated by industrialized countries. China, as the largest developing country in the world, should receive more financial aid from these industrialized countries to finance its low-carbon growth. Given the current global political and economic situation, China must inevitably finance its low-carbon development domestically. China remains in the initial stages of applying fiscal, financial, and market instruments for the benefit of low-carbon growth and is in need of more opportunities to learn, explore, and draw on lessons from abroad. The nexus between climate policy, fiscal policy, and taxation policy needs to be enhanced and made more responsive to the trends of technical advancement and industrial development. Some innovative financial mechanisms and international experiences need to be refitted to be applicable locally, and Chinese financial institutions must find their own sustainable business models that support low-carbon growth.

II The Status of Low-Carbon Financing in China and Abroad

A *The Status of Low-Carbon Financing in China*

Since the great growth experienced by Chinese economy as a result of the liberalization and reform policies, there has been ever increasing pressure on the environment and resources, imploring China to find a sustainable way to develop its economy. In 2003, China began regarding its development through a scientific lens, and has since used this perspective to guide its economic and social development. In June 2007, a group of leaders was set up in the State Council with Premier Wen as head to address climate change actions, energy conservation, and emissions reduction. China developed a "National Proposal on Climate Change Mitigation," and has drafted and revised a number of laws and regulations to follow, such as the *Law on Energy Conservation*, the *Renewable Energy Law*, the *Circular Economy Promotion Law*, the *Cleaner Production Promotion Law* and the *Ordinance on Residential Building Energy Conservation*. These commitments and actions demonstrate the central government's intention to develop the low-carbon economy push toward low-carbon financing.

1 Public Fiscal Support for Low-Carbon Development

Public financing refers to both the policies that center on the general objective of low-carbon growth and to the specific indicators related to energy conservation and emissions reduction set by central and local governments. It further includes official financing activities intended for low-carbon development using budget allocations, fiscal subsidies, transfer payments, government incentives, government procurement, preferential taxes, etc.

China has unveiled a suite of public finance policies that promote energy conservation and emissions reduction over the last few years. China has established a fund dedicated to reorganizing and closing small coal mines, a centrally-financed bonus fund for eliminating old generators, and a subsidy fund for industrial enterprises' energy management centers to preform demonstration projects. The Chinese government has also established tax breaks, investment tax compensation, and accelerated depreciation reimbursement for energy-saving and consumption-reducing enterprises.

To stimulate energy-efficient building reconstruction, the government has created a fund dedicated to reconstructing government and large public buildings, a bonus fund for encouraging the refurbishing of residential buildings in northern China which are centrally heated, and a fund dedicated to renewable energy applications in existing buildings. The government has launched a special subsidy dedicated to low-carbon transportation available to a thousand enterprises and a special fund for transport-oriented energy conservation and emissions reduction efforts.

China has also implemented a number of policy measures designed to encourage the development and integration of low-carbon consumer goods. These include an official procurement plan for energy-efficient products, promotion of energy-efficient products, an old-for-new replacement program for automobiles and electronic household appliances, a purchase tax rebate for small, energy-efficient vehicles, peak pricing for residential electricity use, removal of reduced electricity prices for enterprises with high energy consumption, and an increase to non-residential on-grid power prices in some provinces and cities.

Apart from promoting energy conservation and curbing energy consumption, public finance has also focused on promoting new and renewable energy development. Tax breaks intended to further this objective include: a tax break for the projects listed in the Catalog Guide for Renewable Energy Industry Development, a 50% value-added tax (VAT) refund for the wind power industry and a 100% VAT refund for power generated from straw and other wastes effective immediately, and an import tariff VAT exemption for personal use equipment taken out of the total investment amount (excluding the commodities listed in the Catalog of Non-Tax Deductible Commodities Imported for Domestically-Invested Projects), which is meant for domestic investments in hydropower, heat-and-power cogeneration, solar energy, oceanic energy, biomass energy, and wind power.

Special funds dedicated to renewable energy development include: the central government's special fund for wind power industrialization active 2008–2010, which replaces the wind power bonus with a subsidy; a fiscal subsidy for the Golden Sun pilot projects; a fiscal subsidy for solar photovoltaic (PV) building applications; a subsidy for conversion of straw to energy; a subsidy for the construction of "green energy" pilot counties; and a subsidy for cities (or counties) with pilot renewable energy building applications. China has also offset renewable energy generated power feed in tariff by levying an additional tariff on renewable energy generated power, and has structured wind power feed in tariffs such that the benchmark is determined by four resource area categories. China has made a nationwide, uniform, benchmark feed-in tariff for PV power and a fixed tariff for biomass power.

Public financing directly injects greater investments into low-carbon industries to promote the low-carbon transformation of the economy, but more importantly, it send clear signals to the market on China's policy positions, which guide and drive further social capital into the low-carbon economy. During the 11th Five-Year Plan period alone, more than RMB 220 billion in public finance directly invested in energy conservation and emissions reduction. Although it is difficult to calculate the exact amount of social capital that was invested as a result of fiscal investment and financial and tax policies, the achievements made in energy conservation, emissions reduction, and low-carbon industries during the 11th Five-Year Plan period nevertheless demonstrate the critical role of public financing in low-carbon development.

2 China's Domestic Financial Institutions' Low-Carbon Financing Practices

Bank loans serve as the primary source of low-carbon financing. The Chinese government has made enormous efforts to encourage and motivate banks and other financial institutions to offer loans for low-carbon development and has been consistently optimizing its financial strategy. Using the *Law on Energy Conservation*, the *Renewable Energy Law,* and *the Circular Economy Promotion Law*, the government had expressed that financial institutions should provide preferential loan agreements to qualified energy-conservation and renewable energy projects. The People's Bank of China (PBoC) and the China Banking Regulatory Commission (CBRC) have established relevant policies to meet this demand.

In 2007, the PBoC issued the *Guidelines for Improving and Strengthening Financial Services in Field of Energy Conservation and Environmental Protection* to facilitate the implementation of China's national energy conservation and emissions reduction strategy. In the same year, the CBRC published the *Guidelines for Energy Conservation and Emissions Reduction Credit*, which required financial institutions to accept the strategic importance of financial services in the field of energy conservation and environmental protection, strengthen the coordination and cooperation between credit policies and the general objective of building an energy efficient and environmentally friendly society, and actively contact environmental protection administrations in establishing green credit mechanisms.

To ensure the 11th Five-Year Plan energy conservation and emissions reduction targets are fulfilled, the PBoC and the CBRC jointly issued the *Opinions on Further Improving Financial Services to Support Energy Conservation, Emissions Reduction, and Elimination of Obsolete Mechanisms* in 2010, requiring financial institutions to improve various aspects of their financial services and to take proactive measures to establish and improve a long-term mechanism that supports energy conservation and elimination of obsolete mechanism. To meet the demands of the State Council's comprehensive national macro policies for energy conservation and emissions reduction in the 12th Five-Year Plan, the CBRC issued the *Notice on Releasing Green Credit Guideline* in 2012, requiring financial institutions to strategically promote green credit and to offer greater support for the green economy, low-carbon economy, and circular economy.

Guided by the central government's financial policies, the Chinese banking industry has strongly boosted the availability of "green" credit, sharply decreased loans for carbon- or energy-intensive industries, and eliminated projects that did not conform with the state's policies on energy conservation, emissions reduction, and environmental protection by improving its environmentally related credit policies. The industry has further refined its credit policies dealing with energy conservation and emissions reduction, offering more loans to "green" and low-carbon industries, and focusing on projects with high long-term industrial growth potential. Chinese financial institutions provide crucial support for hydropower, wind power and other new energy industries, as well as for energy conservation projects and equipment manufacturing for industry, transportation, power grids, construction, and household appliances. It actively bolsters industries with outstanding emissions reduction benefits, such as municipal domestic sewage and industrial sewage treatment.

In 2011, the banking industry reduced its balance of credit loans for industries producing surplus capacity by 0.14% year-on-year (YoY) and supported 28.79% YoY more projects related to energy conservation and environmental protection. The industry increased its balance of credit loans for energy-conserving and environmentally-friendly projects by 25.24% YoY by providing bank loans, issuing short-term bonds, mid-term bills, and other financial instruments, pledging receivable accounts, offering anticipated profits, providing equity, and offering factoring services.[1]

The Chinese banking industry has experimented with various low-carbon financing practices. For example, China Development Bank has proactively explored green financial services, annually increasing its loans issued for energy conservation, emissions reduction, and environmental protection, going from RMB 69.3 billion in 2007 to RMB 228.1 billion in 2011. The Bank's 2011 year-end low-carbon credit balance summed to RMB 658.3 billion.[2] The Industrial and Commercial Bank of China has used scientific management to optimize its credit resource allocation and to strictly control the credit threshold for industries with high pollution, high energy consumption, or surplus capacity, using credit leverage to promote China's low-carbon industrial structure. As of 2011, the balance of loans issued for energy conservation, emissions reduction, and clean energy projects amounted to RMB 507.4 billion.[3] SocGen

1 *2011 Report on the Social Responsibility of China's Banking Industry*, issued by China Banking Association.

2 *2011 Report on the Social Responsibility of China Development Bank.*

3 *2011 Report on the Social Responsibility of Industrial and Commercial Bank of China.*

has set up an in-house department dedicated to low-carbon financing and has developed diversified and distinctive financial products in accordance with the emissions reduction market demand.

Not only banks, but also financial institutions engaged in trusts, insurance, and securities have played an active role in promoting low-carbon financing.

3 Carbon Market Pilots and Further Exploration

Since the Kyoto Protocol came into effect in 2005, China's carbon trading market has been dominated by CDM projects and the Voluntary Emission Reductions (VER) market. As of September 13, 2012, NDRC has approved 4,680 CDM projects. By October 17, China had successfully registered 2,431 projects with the United Nations CDM Executive Board (EB), which accounted for 50.89% of all projects registered by host countries. The annual emissions reductions produced by these projects are expected to reach 433,661,991 tons of CO_2 equivalent, which would account for 64.90% of the total annual emissions reductions from projects registered by host countries. As of October 9, 2012, a total of 608,288,832 tons of CERs from China's 930 CDM projects have been issued, accounting for 59.92% of CERs issued for CDM projects by host countries. According to a conservative estimate, these CDM projects have raised 3 to 6 billion USD for China's low-carbon development. In this way, CDM projects have provided many low-carbon financing opportunities for China's enterprises.

Environment Exchanges in Beijing, Shanghai, and Tianjin, have actively engaged in the VER trading market. Enterprises purchased VERs to demonstrate social responsibility. Due to lack of rigid demand, the transaction volume on the VER market is very low. On June 13, 2012, NDRC officially released the "Interim Measures on the Administration of Greenhouse Gas Emission Reductions Transactions," in which VER projects, emission reductions, VER transactions, and verification methods are defined. These "Measures" will play an important role in the development of the VER market, but cannot be expected to change the supply and demand imbalance on the VER market.

The *Communist Party of China (CPC) Central Committee Advice for the 12th Five-Year Plan on National Economic and Social Development* established a clear commitment to "gradually promote the carbon trading market" in October 2010. In March 2011, the 12th Five-Year Plan stated that China would launch low-carbon pilot projects and gradually establish the carbon trading market. Following this announcement, the *Comprehensive Implementation Proposal of the 12th Five-Year Plan on GHG Emissions Control* further clarified specific tasks such as launching carbon trading pilots projects and strengthening the development of auxiliary systems for carbon trading.

The National Development and Reform Commission (NDRC) issued a "Notice on Carrying out Carbon Exchange Pilots" in November 2011, building on the government's declaration, and launching the carbon exchange pilots in Beijing, Tianjin, Shanghai, Chongqing, Hubei Province, Guangdong Province, and Shenzhen. The seven carbon exchange pilot regions are presently working on developing basic emissions data, corresponding technologies, market rules, and carbon exchange administration platforms. They are also studying issues in detail, looking at specific proposals for emissions credit allocation, a practical measurement standard, reporting and verification (MRV) systems, a registration system, trading rules, etc. As of August 2012, Beijing has presented its pilot proposal to the NDRC for approval; Shanghai has finished the allocation of the first group of carbon emissions credits and is conducting polling among corporate entities; and Hubei Province has finalized the fifth version of its carbon exchange administration plan, which specifies the exchange administrator, the exchange supervisor, principles of credit allocation and regulation, and prices for credits.

The carbon trading market has demonstrated huge potential in China. China has participated in global carbon trading mainly through CDM projects in recent years. As the largest carbon emitter in the world, China will probably be required to undertake mandatory emissions reduction obligations after 2020 similar to those adopted by industrialized countries. The Chinese carbon market has to evolve from a CDM and voluntary carbon trading market into a nationwide, standardized market, using incremental cap-and-trade as the primary transaction determinant, and with prospects of linking to the global carbon market.

4 The Current Status of Capital Market Financing

Capital markets are an important channel for directly financing the low-carbon industry, and in reciprocation, the development of low-carbon economy also boosts the capital market and stimulate innovation.

The stock market is an important component of the capital market, and can be divided into the main board market and the second board market. The threshold for listings on the main board remains very high, and stocks issuance is subject to various restrictions. Low-carbon related enterprises will struggle to raise large amounts of capital in the main board market, and emerging small and medium sized low-carbon energy enterprises face particularly extreme difficulty in receiving direct financing under the current mechanism. For example, energy management contracts (EMCOs) were introduced into China in 1990s, and have undergone rapid development, promising a huge potential market. But EMCOs are deterred from the main board market because of its high barriers of entry.

In comparison, the second board market enjoys more flexible mechanisms and barriers of entry, rendering it more suitable to provide precious financing channels for low-carbon enterprises. Although China's stock market was depressed in 2012, many new energy, energy conservation, and environmental sector enterprises listed on the second board demonstrated strong growth. It is expected that when the market realizes the full potential of their profitability in the near future, the second board market will provide crucial financial support for emerging low-carbon enterprises.

The bond market is another important financing channel within the capital market. Financing through bond market remains extremely underdeveloped and stagnant. Because bond financing requires strict criteria, high issuing costs, and long issuing time, low-carbon enterprises or only just beginning to use the bond market to secure financing. This financing channel requires urgent reform.

Venture capital (VC) and private equity (PE) are an indispensable source of financing for the emerging low-carbon industry. China's VC and PE markets have undergone rapid development in recent years. The government has gradually eased legal and regulatory restrictions in order to broaden financing sources. According to China Venture statistics, since 2009, the scale of new energy, energy conservation, and environmental industry financing is experiencing annual growth. In 2009, 73 investment cases were disclosed, with investments totaling 724 million US dollars. In 2010, there were 96 investment cases, totaling 1.289 billion US dollars. In the first half of 2011, 35 investment cases were disclosed totaling 1.039 billion US dollars. China's VC market remains dormant compared with those is developed countries, but it has nevertheless boosted China's low-carbon development.

B *Lessons from International Low-Carbon Financing Practices*

European and US governments and financial institutions, including commercial banks, investment banks, funds, and insurance and venture capital institutions, have all been participating in financing low-carbon markets. It is imperative that the Chinese low-carbon economy learn from international experiences and connect to the international market.

1 Promotion of Low-Carbon Development through Public Financing

Public financing serves as a cornerstone of low-carbon economic growth promotion in industrialized countries. Their public financing policies fall into two major categories. The first category is driving the development of low-carbon industries by encouraging market investment in energy efficiency, energy-related research, and R&D, and developing new renewable energy through fiscal subsidies, budget appropriations, tax deductions, loan discounts, and other similar measures. The second category is limiting the production and consumption of carbon, increasing fossil fuel energy production costs, and implementing measures for energy conservation and GHG emissions controls such as energy and carbon taxes.

Since the 1970s, industrial countries have preferred fiscal subsidies and loan discounts to encourage enterprises to invest in energy efficiency. For example, Australia's *GHG Reduction Scheme* concentrates its subsidies on projects that produce annual reductions greater than 250 thousand tons of carbon dioxide equivalents. The Dutch government provides small and medium sized enterprises (SMEs) with a dedicated energy efficiency investment subsidy policy, which provides funding for 25% of costs incurred in R&D of heat recovery, heat pumping, and absorption cooling technologies.

Budget appropriations are another important source of funding for low-carbon finance. The UK includes a "carbon budget" directly into its overall fiscal budget and invests large amounts of funds into energy efficiency R&D and demonstration projects every year. To support the implementation of the "carbon budget," the UK government set aside GBP 1.4 billion from its overall budget in 2009 for direct investment in sectors with low-carbon activity. It also plans to provide domestic energy technology developers with GBP 5.5 billion of funding for R&D of energy efficiency, disbursed incrementally over the next decade.

Preferential taxation is also widely used in industrial countries for low-carbon financing. In Canada, tax reported depreciation of corporate equipment dedicated to energy efficiency improvement and renewable energy development can be accrued 30% faster. In Japan, a certain percentage of domestic companies' energy efficiency improvements purchase costs can be deducted from the companies' income taxes. In the *Energy Investment Deduction Plan*, The Netherlands has declared that half of the cost of energy-conserving equipment purchased by a domestic company can be deducted from its pre-tax corporate profits that year.

Finally, energy and carbon taxes were first imposed in North European countries and later promoted in other European countries as well. European countries levying energy, carbon, or other similar taxes include Denmark, Finland, Norway, Sweden, Germany, Italy, Switzerland, The Netherlands, the Czech Republic, Austria, Estonia, and the UK. Energy taxes are usually applied to fuel oil, liquefied petroleum gas (LPG), and natural gas.

2 Low-Carbon Financial Products and Innovative Financial Services

Following the establishment of measures to develop low-carbon economies, foreign financial institutions have facilitated the expansion of low-carbon businesses by innovating business operation models and risk control methods. International commercial banks are evaluating the impact of credit projects while operating their credit business, emphasizing environmental risk monitoring in the crediting process, and offering more loans to low-carbon projects.

A variety of innovative financial products related to climate change have been developed. Foreign financial institutions have used their flexibility to spread their investments across Certified Emissions Reductions (CERs) under the CDM, and across European Union Allowances (EUAs) under the EU Emissions Trading Scheme (EU-ETS), to develop various arbitrage products that link different markets, such as CERs and EUAs trading as well as CERs and Emissions Reduction Units (ERUs, under the Kyoto Protocol) trading. These financial services reduce climate change risks and improve the efficiency of the pricing mechanism's allocation of resources to the sectors with cleaner production processes.

3 Rapid Development of the International Emissions Trading Market

Since the initiation of the Kyoto Protocol (KP) in February 2005, the international emissions trading market has boomed, enjoying a constant increase of participants and multiplying business volumes. The KP market consists predominantly of the ETS, the CDM, and Joint Implementation (JI) markets. Major non-KP markets include the US voluntary carbon markets (in some states), Australia's mandatory emissions reduction scheme, the Chicago Climate Exchange (CCX) in the US, and the retail markets formed by internal trading schemes within some corporations (such as BP and Shell).

The total value of the international carbon market grew by 11%, to USD 176 billion in 2011, and the market covered a record high of 10.3 billion tons of carbon dioxide-equivalent. According to a forecast by the United Nations and the World Bank, global carbon exchange volumes will reach USD 150 billion by 2012.

Industrial countries have begun to develop a variety of low-carbon financing systems supported by a number of financial instruments including direct investment and financing, bank loans, carbon funds, carbon credit trading, and carbon futures and options trading.

III Economic Loss and Capital Demand in China's Low-Carbon Development

Emissions reduction projects and the construction of corresponding infrastructure require enormous investment, leaving much room for further low-carbon development. A mechanism is needed to support the supply and flow of tremendous amounts of capital that fund low-carbon economic growth. A recent report from the World Bank notes a huge capital shortfall that hinders climate change adaptation and emissions reduction activities in developing countries. According to the Stern Report, the globe can avoid future annual economic loss equivalent to 5–20% of its GDP by investing 1% of its annual GDP in the low-carbon economy now.

China needs to invest large amounts of money into technology and equipment to further low-carbon development. The investment will make a prominent contribution to the development of non-fossil energies and the promotion of efficient energy technologies, but it will also bring result in economic losses. China must analyze how much investment is needed and how much GDP loss will result if a low-carbon model is adopted for China's economic growth. This paper calculates GDP loss possibly incurred by China's investment in low-carbon development using the computable general equilibrium (CGE) model, and goes on to consider the investment required to achieve China's climate change goals using a bottom-up model.

A *GDP Loss Resulting from Low-Carbon Development*

This paper uses the CGE model to estimate the costs and incremental investment needed for low-carbon development by analyzing two different scenarios: the business-as-usual (BAU) scenario and the reduced emissions scenario.

1 The BAU scenario

The BAU scenario assumes no new policies on emissions reduction or any new GHG emissions reducing technologies, and the Compound Annual Growth Rate (CAGR) of GDP is assumed to be 7.8% from 2011 to 2015 and 6.5% from 2015 to 2020. Under this scenario, China's demand for primary energy will grow 6.6% annually from 2010 to 2020, so that total energy consumption will increase from 2.359 billion tons in 2005 to 4.501 billion tons in 2015 and further to 6.2 billion tons in 2020. Coal and petroleum will continue to serve as the pillars of primary energy consumption. The total power generated from coal and petroleum will grow in volume but shrink in terms of energy share, while the demand for natural gas, nuclear power and non-hydro renewable energy will increase.

2 The Reduced Emissions Scenario

Under this scenario, China makes significant contributions to global efforts against climate change by enacting measures across sectors to control total energy consumption and optimize the energy consumption structure. Demand for primary energy will grow 5.7% annually from 2010 to 2020, so that total energy consumption will increase from 2.359 billion tons in 2005 to 4.1 billion tons in 2015 and further to 5.4 billion tons in 2020. From 2011 to 2020, a series emissions reduction policies will be introduced, mature low-carbon and emissions reduction technologies will be promoted and applied, many advanced technologies (such as energy efficiency technologies and some renewable energy technologies) will be adopted, and investment will be injected into technical R&D for carbon capture and storage (CCS), solar power generation, electric vehicles.

Figure 9.1 shows the GDP loss incurred by emissions reduction activities calculated using the CGE model for this scenario, estimating a 1.5% loss of total GDP by 2015 and a 2.47% loss of total GDP by 2020.

B *Demand for Incremental Investment in Low-Carbon Development*

Many studies have estimated the capital investment demanded for China's low-carbon growth using different low-carbon scenarios and yielding different results. All studies agree that China's low-carbon growth demands a tremendous amount of incremental investment to the energy sector and to end-users.

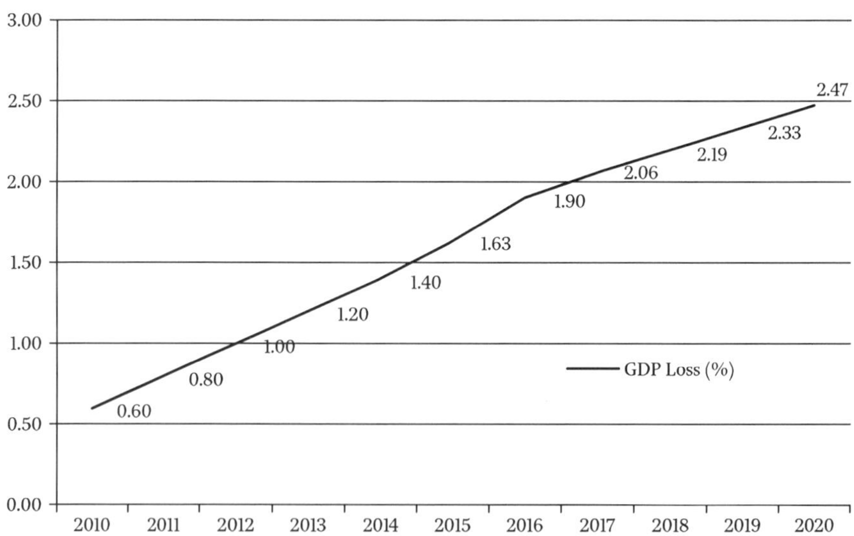

FIGURE 9.1 *GDP Loss under the Reduced Emissions Scenario (%)*

This paper calculates the incremental investment needed to realize low-carbon development by observing investment in major sectors such as energy, industry, transportation, and construction, based on the goals described in the 12th Five-Year Plan.

TABLE 9.1 *Investment demanded for low-carbon growth in different sectors during the 12th Five-Year Plan Period*

Sector	Focus of the sector's emissions reduction activities	Energy conserved (million tons of coal equivalent)	Investment (RMB trillion)
Energy	Power generation and distribution technologies, new energy technologies.	264[4]	2.46[5]
Industry	By 2015, energy consumption per unit of industrial value added is to be reduced by (compared with 2010 levels): 18% for steel, 18% for non-ferrous metallurgy, 18% for petrochemicals, 20% for chemicals, 20% for construction materials, 22% for machinery, 20% for light industry, 20% for textiles and 18% for electronic information industries.	670[6]	1.68[7]
Transportation	According to the 12th Five-Year Plan, the overall energy consumption per unit of railway transport volume is to be reduced	1[8]	0.05[9]

4 In 2015, the power industry will save 264 million tons of standard coal equivalent compared with 2010 levels.

5 The investment includes RMB 660 billion for power generation and distribution technologies and RMB 1.8 trillion for new energy technologies, assuming a balance between investment and returns for energy conservation.

6 The 12th Five-Year Plan on Industrial Energy Conservation.

7 Assuming a balance between costs and returns for energy conservation in the industry sector.

8 The 12th Five-Year Plan on Energy Conservation and Emissions reduction.

9 In 2011, the government invested RMB 250 million and stimulated market investment of RMB 8.06 billion into the transport industry, with a ratio of 1:32. This analysis assumes that the ratio remains unchanged in the next four years.

TABLE 9.1 (*cont.*)

Sector	Focus of the sector's emissions reduction activities	Energy conserved (million tons of coal equivalent)	Investment (RMB trillion)
	from 5.01 tons of standard coal equivalent per million ton-km in 2010 to 4.76 tons of standard coal equivalent per million ton-km in 2015. Energy consumption per unit of road transport volume is to be reduced from 5.01 kg standard coal equivalent per hundred ton-km in 2010 to 4.76 kg standard coal equivalent per hundred ton-km in 2015. Energy consumption per unit of shipping volume is to be reduced from 6.99 kg standard coal equivalent per thousand ton-km in 2010 to 6.29 kg standard coal equivalent per thousand ton-km in 2015. Energy consumption per unit of air transport volume is to be reduced from 0.45 kg standard coal equivalent per ton-km in 2010 to 0.428 kg standard coal equivalent per ton-km in 2015.		
Construction	By the end of the 12th Five-Year Plan period, the construction industry is to reach an energy conservation capacity	1.16[10]	0.34[11]

10 The 12th Five-Year Plan on Energy Conservation Applications in Buildings.

11 This estimate covers five elements: new constructions, heat-accounting reconstruction, existing building reconstruction in northern China, establishment of a supervisory system, and conventional energy substitution. The incremental cost of energy conservation reconstruction per unit of existing building area is RMB 80–120/m^2, and 400 million m^2 of existing buildings are to be reconstructed for energy conservation purposes during the 12th Five-Year Plan period, resulting in a total cost of RMB 40 billion. Given an incremental cost of RMB 30/m^2 for energy conservation purposes, the total cost will amount to RMB 300 billion as 2 billion m^2 of new construction is to be built during the same period.

Sector	Focus of the sector's emissions reduction activities	Energy conserved (million tons of coal equivalent)	Investment (RMB trillion)
	of 116 million tons of standard coal equivalent. Specifically, 45 million tons will be conserved through green buildings and energy-efficiency applications in new constructions; 27 million tons will be conserved by further reform to heat supply systems, using popular adoption of the accounting-and-charging system and the reconstruction of existing centrally heated buildings for heat accounting and energy conservation purposes in northern China; 14 million tons will be conserved through the establishment of a stronger supervisory system for energy conservation of public buildings and from energy conservation oriented rebuilding and operations management; and the other 30 million tons of standard coal equivalent represents substitution of conventional energy capacity, stemming from the integrated application of renewable energy in buildings.		
Total		10.51	4.53

In the energy sector, power generators require a total investment of RMB 660 billion for energy conservation purposes during the 12th Five-Year Plan, accounting for the investment and return balance of energy conservation in power generation and distribution. According to the 12th Five-Year Plan on Renewable Energy Development, RMB 1.8 trillion is required, leading to a total investment of RMB 2.46 trillion in the energy sector for energy conservation and emissions reduction.

According to the 12th Five-Year Plan on Energy Conservation, the industrial sector is expected to conserve a total of 670 million tons of standard coal equivalent during this period. Accounting for the investment and returns balance of energy conservation in this sector, a total investment of RMB 1.68 trillion is required.

Emissions reductions in the transportation sector cover railway, road, shipping, and air transport segments. According to the 12th Five-Year Plan, a total of 1 million tons of standard coal equivalent should be conserved during the 12th Five-Year Plan period. According to the capital invested by the government in 2011, a total investment of RMB 50 billion is needed for emissions reduction purposes during the 12th Five-Year Plan period.

In the low-carbon construction sector, energy conservation work will focus on new constructions, heat accounting reconstruction, existing building reconstruction in northern China, and establishment of a supervisory system. During the 12th Five-Year Plan period, 116 million tons of standard coal equivalent will be conserved, requiring a total investment of RMB 340 billion, equal to RMB 2,900 per ton of standard coal equivalent conserved.

To achieve its 2015 targets on energy consumption per unit of GDP, energy conservation, emissions reduction, China will need to make a total low-carbon investment of approximately RMB 4.53 trillion from 2010–2015, broken down into an annual investment of RMB 910 billion, which is almost equal to 2.2% of total GDP.

Studies suggest that as energy efficiency measures advance, energy conservation efforts will offer ever decreasing potential but require increasing volumes of investment. Most researchers anticipate a much larger investment in energy conservation during 2016–2020 than that made during 2011–2015. For instance, McKinsey expects the investment in energy conservation in 2016–2020 to be quadruple that of 2011–2015. In 2009, Renmin University of China estimated the annual incremental investment after 2030 will be 1.5 times greater than during 2010–2030. Based on the findings of various institutions and the investment sum proposed in this paper for the 12th Five-Year Plan period, we can estimate that during 2016–2020, China will need a total incremental investment of RMB 6.8–18.12 trillion, or an annual incremental investment of approximately RMB 1.36–3.6 trillion.

IV Some Thoughts on China's Low-Carbon Financing Policy

In conclusion, China has provided comprehensive support for low-carbon growth using fiscal investment, preferential taxation, and financial policy, and has facilitated and achieved significant progress in reducing domestic emissions and developing low-carbon industries. However, low-carbon financing in China still faces many challenges so that there is great need for further work and policy development.

A *Quickly Developing a Strategy for Low-Carbon Public Financing and Leveraging Public Financing*

China must develop a comprehensive mid- to long-term national financial plan for low-carbon growth, explicitly define its strategy objectives and focus on financial policy support. First, the national climate change policy roadmap should be based on a low-carbon finance strategy. Existing low-carbon plans regarding energy, the circular economy, energy conservation, and emissions reduction should be strategically integrated into a national proposal and low-carbon economy action roadmap as a top priority. Such an integrated strategy may serve as a blueprint for low-carbon growth. Second, the government should set mid- to long-term low-carbon financial policy objectives. By integrating both past and current policies based on full scientific analysis, the Chinese government can establish a complete and fully integrated policy system which includes not only a policy strategy for national economic growth but also specific auxiliary policies regarding various sectors and industries, emphasizing both long-run development and mid-to-near-term results.

1. China should rely more heavily on financial policies leverage to achieve its carbon reduction goals. Financial investment and fiscal subsidies will serve as the primary drivers of low-carbon development for the foreseeable future. It is particularly important to select the appropriate projects and improve the efficiency of investments and fiscal subsidies. For energy producers, technical innovation is pivotal in achieving low-carbon growth. Investments should concentrate on R&D, especially demonstration and promotion stages, while fiscal subsidies should focus on the demand side of development, prioritizing low-carbon products in government procurement, continuing feed-in tariffs for wind and solar PV power, and offering subsidies for the purchase of energy-efficient products.

As one of the most effective financial policy instruments for economic regulation, preferential taxation should remain under constant revision by the government to better facilitate the integrated use of resources. A higher percentage should be deducted from the pre-tax expenses incurred in renewable energy R&D; letting taxation policies more actively and effectively guide the efficient development of renewable energy. New energy investment projects with visible potential economic and social benefits can best be promoted with preferential policies such as accelerated depreciation and income tax credits. China can also learn from industrialized countries experiences creating climate change taxation regimes. China should promote the establishment of a diversified and socialized mechanism for low-carbon financing by guiding major energy consumers to sign up for voluntarily energy conservation and emissions reduction agreements.

China should set up a green tax system. First, the government should further resource tax reform by increasing taxes on fossil resource products and broadening the implementation of price-based resource taxation. Second, China should consider accelerating the implementation timeframes for environmental carbon taxes. Once environmental taxes are implemented, some pilot regions may be considered for a partial reform that would charge for pollution discharge, levying taxes pollutants. If environmental taxation is to be implemented nationwide, China should consider a comprehensive reform under which all pollutant based charges are worked into environmental tax. Third, other taxes related to low-carbon growth (such as VAT, consumer taxes, corporate income taxes, and vehicle and vessel taxes) should be adjusted to encourage more investment into low-carbon industries by narrowing the gap in production cost between low-carbon industries and high energy consuming industries.

B *Establishing an Effective Credit Mechanism for Low-Carbon Growth and Innovating Financial Products*

China should establish an effective credit mechanism that supports low-carbon growth. The banking industry has made proactive attempts to make "green" credit available, prompted by recent national policies, achieving considerable economic and social benefits. Chinese banks' low-carbon financing activities remain mostly motivated by national policies and a self-imposed corporate social responsibility. China needs a business model that creates profit incentives for banks investments in low-carbon sectors to remain sustainable in the long-term.

The PBoC, China's central bank, should provide compensation, such as financial discounts, to partially offset the risks commercial banks take when they extend credit to low-carbon projects amid China's changing energy consumption structure and strong economic growth. PBoC should also set up a monitoring system for the credit risks associated with low-carbon-related industries, improve information exchange, and regularly announce project progress to keep commercial banks informed of the status of capital usage and risks and to protect the interests of commercial banks and the security of the banking system.

Policy banks and commercial banks should strengthen cooperation and exploit their respective advantages to provide one-stop financial services for key low-carbon economic projects that require large amounts of starting capital and a long payback period using joint loans, syndicated sub-loans, etc.

Finally, China should establish a mechanism that encourages innovation in financial products. Low-carbon financing does not only require a higher availability of "green" banking credit but also innovation in carbon-finance services. Commercial banks could initiate carbon-credit-pledged loans, carbon-credit-based financial leases, and carbon-credit-based factoring services. They can further work proactively to establish trust in carbon finance products and new financial products such as carbon funds.

C *Accelerating Multi-Level Construction of the Capital Market and Improving Venture Capital Investments*

China should accelerate multi-level construction of the capital market, establishing a market-oriented innovation mechanism to improve market efficiency and further expand the width and depth of the domestic capital market. Specifically, China should:

1) Encourage and assist financing of qualified low-carbon enterprises by issuing shares and prompt listed low-carbon companies to finance themselves through additional share offerings and share allotments on the main board, on the condition that securities regulators do not lower the threshold share for market entry. China should provide rapid growth and expansion support for outstanding enterprises in low-carbon industries by encouraging mergers and acquisitions (M&A). China should also reinforce construction of the growth enterprise market (GEM) and appropriately lower the initial public offering (IPO) standard for small-to-medium-sized new energy enterprises featuring new concepts, fast growth, and warm reception from investors, thus creating a sustainable source of capital for these enterprises.
2) Improve the quality of publicly listed low-carbon companies, and promote their standard operations. The quality of publicly listed low-carbon companies is the source of stock market investment value. The market administrator should effectively regulate low-carbon companies with relatively high risks, standardizing their operations and regularly disclosing their information in order to improve their source quality and strengthen investors' confidence.

3) Promote a faster coordinated development of the corporate bond market and the stock market and further encourage innovations in fixed income products and warrant products such as listed low-carbon corporate bonds, warrant corporate bonds, and standardized asset-securitized products.

China should motivate an increase to private capital flows by continuing to relax restrictions on the participation of private capital in low-carbon industries investment finance activities and extending the rights of various independent energy development enterprises. China should encourage qualified private businesses to become legal developers of alternative energy sources such as large-scale shale gas. The government should authorize the development of hydropower and wind power projects on a fair, open, and just basis to encourage private capital participation, and should encourage private businesses to build and operate such projects in joint cooperation with state-owned enterprises using methods such as joint ventures, cooperation, associated operations, project financing, etc. Energy conservation projects should be completely open to private businesses, who should be encouraged to invest in all aspects of the construction and operation of energy conservation projects. The central government should provide comprehensive policy support for private business participation in energy conservation projects through a variety of measures including preferential taxes, financial aid (such as investment guarantees and corporate bonds), investment subsidies, loan discounts, and even partial capital injections.

D *Building the Carbon Market More Quickly*

The construction of the carbon market is a systematic project with high technical requirements and far reaching social involvement. The relatively mature EU-ETS has been operating for almost a decade and continues to be tinkered with. China's national market is new and immature, suffering from low social consciousness of low-carbon growth and weak fundamental capacity. China faces great challenges in constructing a domestic carbon market. It must consider its domestic conditions and harness momentum from it fundamental economic development to promote incremental change.

1 China Should Improve Carbon-Trading Systems and Strengthen Government Regulation

Development of carbon trading in China requires clear legislation that establishes the legitimacy of carbon emissions transactions. On June 30, 2004, the NDRC, the Ministry of Science and Technology, and the Ministry of Foreign

Affairs jointly issued a paper outlining a set of temporary measures to govern the *Management of Clean Development Mechanism Projects*. In June 2012, the NDRC issued the *Temporary Measures on the Management of Voluntary GHG Emissions Reduction Trading* in order to standardize and streamline enterprises' aggressive trading of voluntary emissions reductions.

To establish a national carbon market, China should first develop carbon emissions credit trading laws and regulate the participation of financial institutions in the carbon market to clarify carbon emissions rights and provide carbon emissions pricing with legal ground and security. China must then release normalized rules and systems to ensure standardization throughout the carbon market development process. These rules and systems may cover the market entry system, administrative systems, the information reporting system, and the supervisory system. Given the large number of carbon trading participants, including buyers, vendors, third-party service providers, legislators, regulators, etc., standardization of market rules will help reduce market volatility.

A strong government regulation mechanism is also indispensable to secure and efficient market operations. The operation of European and US carbon markets demonstrate the need for transparent public decision-making processes and judicial review to provide effective security for the operation of markets like EU-ETS and the Regional Greenhouse Gas Initiative (RGGI).

2 China Should Build a Unified Methodology and Transactions Platform

Seven provinces and cities are currently executing pilot proposals for local carbon trading. As the pilot projects progress, coordination problems will emerge if no national coordination mechanism for the seven different designs in adopted. These issues could easily hinder progress in the field of carbon trading throughout China. China must integrate the carbon trading activities currently decentralized across different industries and regions as soon as possible in order to establish unified methodologies and standards for GHG emissions accounting. Specific tasks includes: studying and developing GHG reporting and accounting methodology in key industries and corporations; preparing an online monitoring system for the energy consumption of major corporations and institutions; establishing certification and verification systems; constructing and improving trading platforms; and creating a link between the Voluntary Emissions Reductions (VER) market and the carbon credit market, connections between different industries, and a link between the Chinese carbon market and the global carbon market.

3 China Should Develop Specialized Institutions and Encourage the Participation of Financial Institutions

A developed carbon market cannot operate without specialized service providers. Specialized companies offer complete information about the carbon market and provide a number of services including carbon asset development, carbon project consulting, emissions reduction index brokerage, project financing, carbon portfolio management, and carbon neutralization products.

Financial institutions (including commercial banks, insurance companies, investment banks, and trust companies) play an important role in improving carbon market liquidity, establishing prices, and controlling risks in carbon markets across the globe. Chinese financial institutions are not currently capable of playing this role as they are not yet familiar with the carbon trading rules and the development of carbon-credit-based financial derivatives. Chinese financial institutions, which are mostly profit-driven, have had little motivation to engage in China's still infant carbon trading market and offer only primary and undiversified financial instruments and services. China should encourage financial institutions to participate more fully in carbon trading and to engage the carbon market.

The Chinese carbon market is dwarfed in comparison to the developed EU-ETS market in terms of policy systems and market environment, and it faces many unique obstacles. China cannot simply copy the process model the EU-ETS used for its development. China must rather appropriately guide its domestic financial institutions and carbon-market institutions toward development of diversified financial products and services that better fit the Chinese carbon market. China must gradually explore multi-level investment and financing options for the carbon market by herding financial institutions such as banks, trust companies, rating agencies, and insurance companies to the carbon market to ensure its proper development.

It is crucial for China to engage in low-carbon development now. It must address energy conservation and carbon dioxide emissions objectives and also spur low-carbon development if it is to achieve sustainable economic growth. With the central government's strong support for energy conservation and emissions reduction, and with the development of low-carbon industries, local governments and enterprises have undertaken aggressive actions and have made admirable achievements, notable growing budget appropriation, green credit loans, and investments in many low-carbon industries. China remains undeniably in the initial stages of applying fiscal, financial, and market instruments in support of low-carbon growth and has much room to explore experiences from abroad, learning and drawing lessons from their trials.

(This article was originally published in Chinese in 2012.)

CHAPTER 10

China's Green Climate Fund: Innovation and Experience

A Case Study of the China Green Carbon Foundation

Li Nuyun and Li Jinliang

Abstract

The Green Climate Fund (GCF) was officially founded at the UN Framework Convention on Climate Change (UNFCCC) Durban Conference as a tool to assist developing countries mitigate climate change and adapt to its effects. Durban leaves operations of the GCF undecided, and the international community lacks successful experience to guide interested countries. As a pioneer of China's GCF, the China Green Carbon Foundation (CGCF) is the first nationwide non-profit public foundation dedicated to carbon sequestration, emissions reduction, and addressing climate change. The CGCF is a competent professional organization implementing carbon offsets and carbon neutral projects in China by employing measures such as afforestation projects to sequester carbon and forest protection to reduce carbon emissions. The CGCF has created an integrated public service platform to help enterprises and the public to realize its four main objectives: storing carbon credits, fulfilling social responsibilities, increasing farmers' income, and improving the ecological environment through forestry measures. The CGCF has actively explored ways to innovate its operational model, standard development, project management, carbon trading, scientific research, publicity, and promotion and engages in a novel form of public foundation operations matching enterprise donations for afforestation with corporate social responsibility, carbon credit storage, and voluntary emission reductions. This model has provided a new effective path for businesses and the public to participate in addressing climate change. This paper elaborates on the CGCF innovative practices over the past five years including its operational model, successful projects, and practical experience, and proposes public foundations as

* Dr. Li Nuyun is the Deputy General Director at the Office for Combating Climate Change, State Forestry Administration of China, Secretary General of the China Green Carbon Foundation. Her major areas of study include forestry and climate change, carbon trading and ecosystem services market, and social impact assessment of forestry projects. Dr. Li Jinliang is Chief Engineer at China Green Carbon Foundation. His major areas of study include forestry and climate change, sustainable forest management, and management of forestry projects.

 | DOI 10.1163/9789004274648_011

important actors in combating climate change. It provides effective reference models and experiences for the Durban GCF operations and management, which promote healthy development toward the GCF's set goals.

Keywords

The Green Climate Fund – China Green Carbon Foundation – afforestation – carbon sequestration and emission reduction – addressing climate change

I Introduction

The Green Climate Fund (GCF) is regarded as one of the more important achievements of the seventeenth sessions of the Conference of the Parties (COP 17) to the UN Framework Convention on Climate Change (UNFCCC) and the seventh meeting of the Conference of the Parties that served as the Meeting of Parties of the Kyoto Protocol (CMP 7) held in Durban, South Africa, at the end of November 2011. The GCF is intended to support developing countries acceded to the UNFCCC, in particular aiding countries with the weakest policies and activities take low-carbon measures among others to adapt to climate change. The Durban Conference defined the GCF as the Convention's financial operating entity and required initiation of fund management as soon as possible as well as work plan development to manage long-term funding support for developing countries. The GCF is expected to annually mobilize at least US $100 billion by 2020. The Durban conference decision also urged Convention financial mechanisms such as the GCF to provide developing countries with financial support as soon as possible to reduce emissions from deforestation and other activities. Unfortunately, key issues concerning the GCF such as its operations model and management system were not decided in the Durban Climate Conference, lacking successful international experiences as references, so that further negotiations at upcoming conferences are needed. Given the rarity of reports in China and abroad on successfully operated public funding foundations specifically dedicated to combating climate change, this paper uses the China Green Carbon Foundation (hereinafter referred to as the CGCF)—a pioneer of China's GCF—as a case study. It demonstrates how China spearheaded the operation of the first nationwide non-profit public foundation dedicated to addressing climate change, thereby providing examples of effective models and experiences for the operation and management of the Durban Green Climate Fund.

II Background on the Establishment of China's Green Climate Fund

Global climate change, particularly global warming, is one of the greatest current threats to human society. Addressing climate change has become a worldwide political, economic, diplomatic, and ecological hot-button issue. Using forestry measures to mitigate climate warming is recognized as an effective approach and is actively promoted by the international community. The Fourth Assessment Report by the Intergovernmental Panel on Climate Change (hereinafter referred to as the IPCC) states: "forestry has multiple benefits, including the dual functions of climate change mitigation and adaptation, and has been acknowledged as an important, low-cost, and economically viable measure for the next 30 to 50 years." Forestry measures have hence been increasingly incorporated in international climate change efforts.

According to the Convention's "common but differentiated responsibilities" principle, China is not required to adopt mandatory emissions reduction obligations prescribed under the Kyoto Protocol. As the largest emitter of greenhouse gases worldwide and a responsible major power, China has taken specific actions to become more resource efficient and environmentally friendly, meeting both the needs of the international community and China's long-term development strategy. In accordance with the basic principles of the Convention, the Chinese government is taking proactive measures to reduce greenhouse gas emissions and mitigate global warming. These endeavors encompass energy conservation, consumption reduction, new energy and renewable energy development, as well as a host of other vigorous efforts to increase carbon storage reduce emissions through measures such as afforestation, sustainable forest management, and forest conservation. China has adopted the forestry objective as one of the three major objectives of the voluntary commitments to emissions reduction presented at the UNFCCC conference in Copenhagen in 2009, including committed increases to forest carbon sinks, a 40 million hectare increase to forest area, and a 1.3 billion cubic meters increase to forest standing stock volume by 2020, based on the 2005 levels (hereinafter referred to as the "dual increases" goal).

The Chinese government has always emphasized restoration and conservation of forest vegetation. According to the Seventh National Forest Resources Inventory,[1] China had a forest area totaling 195 million hectares, forest coverage of 20.36%, a total forest tree stocking volume of 14.913 billion cubic meters, and carbon stocks in forest vegetation amounted to 7.811 billion tons.

1 SFA, *China's forest resources report—The Seventh National Forest Resources Inventory*, (Beijing: China Forestry Publishing House, 2009).

Forests provide important ecosystem services such as carbon fixation, oxygen production, water retention, soil conservation, air purification, accumulation of dry matter, and biodiversity conservation, which are worth an annual value of RMB 10.01 trillion yuan. China has become the country with the largest growing forest area in the world, making important contributions to global warming mitigation, winning it acknowledgment and praise from the international community. According to the State of the World's Forests 2010 report[2] released by the UN Food and Agriculture Organization (hereinafter referred to as the FAO), in Asia and in the Pacific region as a whole, forests were lost at a rate of 700,000 hectares per year in the 1990s, but grew by 1.4 million hectares per year over the period of 2000–2010. This reversal was primarily credited to large-scale afforestation efforts in China, where the forest area increased by 2 million hectares per year in the 1990s and by an average of 3 million hectares per year since 2000. Because China is a developing country, the carbon sinks created by the large-scale afforestation could not be used to offset carbon emissions in the way Kyoto Protocol Annex 1 countries use them, but can only serve for positive publicity. In the absence of a greenhouse gas emissions cap set for enterprises by the country, China and Chinese enterprises could reap great benefits if the carbon sinks could be converted to carbon credits using technical measures and funding channels available to enterprises, such as considering them a first deposit on enterprises' carbon credit accounts for low-cost emissions reduction reserves to be used by domestic enterprises in the future. Under the current circumstances, China cannot meet its carbon sink increase and emissions reduction objectives needed to address climate change and at the same time satisfy the demand for ecological products and social development progress by relying solely on the government's efforts to restore and conserve forest vegetation. A public platform to draw donations from enterprises in support of afforestation, forest conservation, and sustainable forest management is needed to supplement. The extra funds could increase forest cover, safeguard national ecological security, and also help enterprises undertake voluntary emissions reduction, participate in the global movement to address climate change at lower costs, build a good corporate social image, enhance soft power, and promote corporate sustainable development. Thanks to donations from the China National Petroleum Corporation (CNPC) and the Sino-Forest Corporation Investment Co. Ltd., the CGCF was established on July 19, 2010 with approval from the State Council, and was registered at the Ministry of Civil Affairs. The CGCF is governed by the State Forestry Administration (SFA). It is the first nationwide non-profit public foundation dedicated to

2 FAO: State of the World's Forests 2010. 2011.

combating climate change by increasing carbon sinks and reducing carbon emissions in China, and it can be regarded as China's first Green Climate Fund. According to the Articles of the CGCF, the scope of the CGCF's work includes: combating climate change through afforestation, forest management, desertification control, development of energy forest bases, wetland protection, and biodiversity conservation; establishing various types of memorial forests for the purpose of accumulating carbon stock and designating green space creation; strengthening the protection of forests and woodlands, and reducing the carbon emissions caused by inappropriate land use; supporting R&D, education, and training for public welfare improvements, carbon sink increases, and emissions reduction; accounting and monitoring of carbon sequestration and formulating relevant standards; actively publicizing the ways in which forests address climate change, and raising public awareness of the importance of protecting the ecological environment and paying attention to climate change; conducting cooperation and exchange activities in China and abroad on the forest's role in combating climate change; and undertaking other appropriate social welfare activities in line with the goals of the CGCF.

III China's Green Climate Fund's Innovations and Practices

Public foundations working in the interest of public welfare by increasing carbon sinks reducing emissions to address climate change are not a common phenomenon in China or abroad, leaving few examples to draw from. The CGCF was given a pioneering mission to explore and innovate in the field, and many tasks are being tried for the first time, requiring great creativity. The CGCF's mission is to "increase green vegetation, absorb carbon dioxide, address climate change, and protect our planet." There have already been a number of experiments and trails testing operation models, standard development, project implementation, carbon trading, scientific research, publicity and education, and management regulation, accumulating significant results

A *Public Service Centered, Innovative Operational Model*

According to the "Regulations for Foundation Management" from the State Council in China, the "Public Welfare Donations Law" of the People's Republic of China, and the "Articles of China Green Carbon Foundation," the CGCF activities are to be carried out in a way resembling all other nonprofit foundations focused on the public interest. One important difference is that aside from the CGCF's public welfare activities, it can also generate and record additional "carbon credits" in enterprise or individual carbon sequestration credit

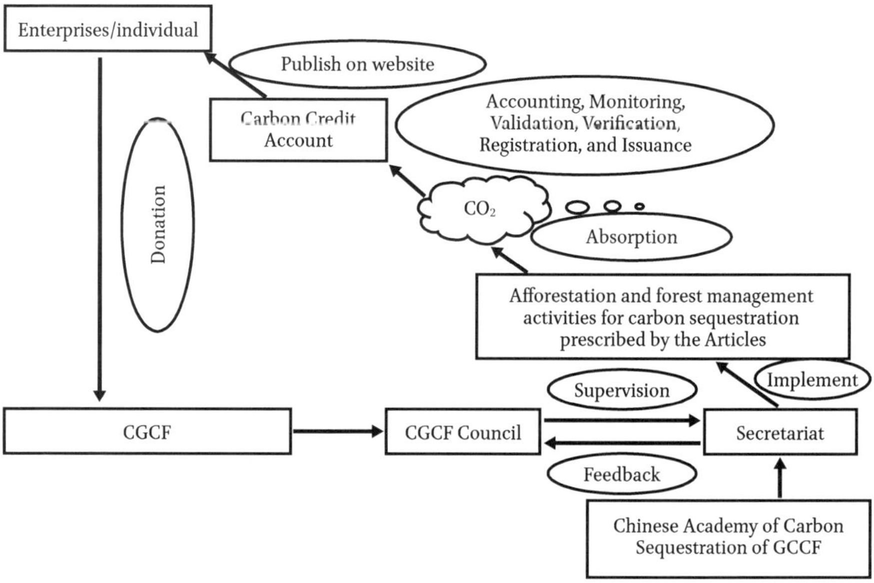

FIGURE 10.1 *The China Green Carbon Foundation Operational Framework*

accounts. This unique innovative model for the CGCF makes it significantly different from other public welfare foundations (see Figure 10.1).

The CGCF has developed a series of rules and regulations including the "CGCF Fund Management Measures" and the "CGCF Project Management Measures." CGCF is committed to building a public service platform addressing four objectives: storing carbon credits, fulfilling social responsibilities, increasing farmers' income, and improving the ecological environment through forestry measures. The CGCF's management and operational follows the following model (see Figure 10.2.) An enterprise or individual makes a donations to the CGCF; the CGCF organizes the specialized organizations and implements afforestation projects to sequester carbon and forest management projects, in accordance with the donor's wishes; the land that is afforested and the forests managed already belong to local farmer's and land owners; farmers in the project region receive employment opportunities and supplement their income by participating in afforestation and forest management activities. The projects provide further benefits including biodiversity conservation, environmental improvements, forest products and byproducts production, and creation of recreational places that contribute to green growth and combat climate change. Donors get carbon credits generated by the project, in accordance with the accounting, monitoring, verification, and registration work done by

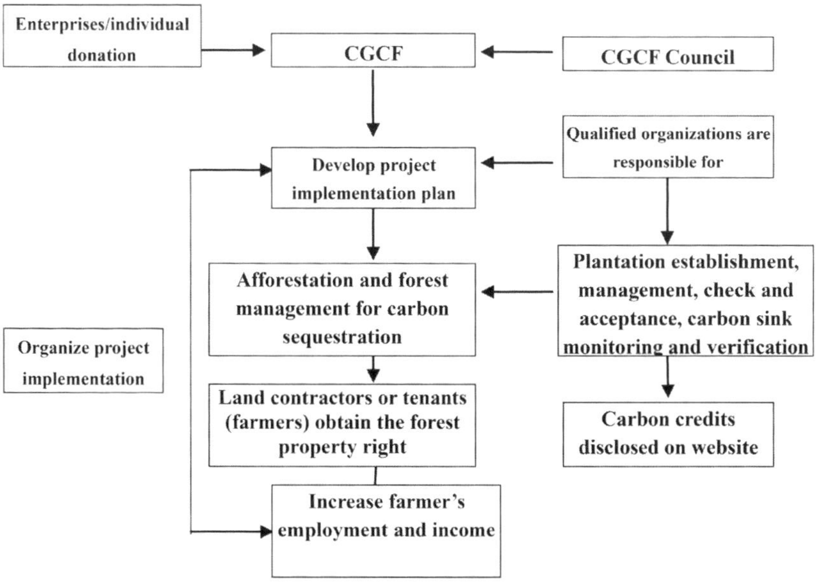

FIGURE 10.2 *The China Green Carbon Foundation Operations Model*

professional organization, and the credits are recorded in the enterprises' or individuals' social accountability accounts and published the on the official CGCF website.

B *Accelerating Study and Development of Standards to Regulate Project Implementation*

In order to regulate project implementation, the CGCF has been busy drafting a series of carbon sequestration project regulations and rules for forestry projects organized by the SFA. These include the "Afforestation Technical Code for Carbon Sequestration" (Trial), the "Verification and Acceptance Measures for Afforestation Carbon Projects" (Trial),[3] the "Guidelines for Carbon Accounting and Monitoring Afforestation Projects"[4] (Released by the State Forestry Administration) which have already been applied in project implementation. The CGCF has organized experts to develop four carbon sequestration methodologies, from which the "Bamboo Afforestation Methodology for

3 SFA: "Carbon Sequestration Afforestation Technical Code" (Trial), and "Verification and Acceptance Measures for Forestry Carbon Projects" (Trial) (BAN ZAO ZI [2010] No. 84号), 2010.

4 SFA: "Guides for Carbon Accounting and Monitoring for Afforestation Projects" (BAN ZAO ZI [2011] No. 18), 2011.

Forest Carbon Projects" has already passed expert panel review and been used in domestic and international bamboo afforestation projects trials for carbon sequestration. The "Validation and Verification Guidelines for Forestry Carbon Projects" (Trial) and "Technical Code for Forest Management Carbon Projects in Wenzhou" have also been applied in project implementation. The "CGCF Interim Management Measures for Forestry Carbon Project Registration" has been formulated, and its supplementary registry system for forest carbon projects has been developed used in operation trials. The CGCF has established a preliminary standard system for forest carbon sequestration projects. which does not only match international standards, but further addresses special circumstances present in China, and has laid a foundation for the scientific and regulated implementation of carbon sequestration afforestation projects (see Table 10.1).

TABLE 10.1 *CGCF Standards and Regulations (Trial)*

Category	Name (Note)
Project Methodology Including (Afforestation and reforestation) (Forest management)	"Arbor Afforestation Methodology for Forest Carbon Projects" (to be examined)
	"Shrub Afforestation Methodology for Forest Carbon Projects" (to be examined)
	"Bamboo Afforestation Methodology for Forest Carbon Projects" (Trial)
	"Sustainable Forest Management Methodology for Forest Carbon Projects" (to be examined)
	"Afforestation Technical Code for Carbon Sequestration (Trial)" (Released by the State Forestry Administration on trial)
	"Validation and Acceptance Measures for Afforestation Projects for Carbon Sequestration (Trial)"
	"Guidelines for Carbon Accounting and Monitoring for Afforestation Projects" (Released by the State Forestry Administration)
	"Preferred Model for China's Carbon Sink Increase and Emissions Reduction through Sustainable Forest Management" (under development)
	"Guidelines for Carbon Accounting and Monitoring of Sustainable Forest Management Projects" (under development)
	"Technical Code for Forest Management Carbon Projects in Wenzhou" (Trial)

Category	Name (Note)
Validation and Transaction	"Validation and Verification Guidelines for Forestry Carbon Sequestration Projects" (Trial) "Forestry Carbon Credit Trading Standards" (Trial) "Forestry Carbon Credit Trading Rules" (Trial) "Forestry Carbon Credit Trading Process" (Trial) "Forestry Carbon Credit Trading Account Escrow Rules" (Trial) "Forestry Carbon Credit Trading Escrow Agreement" (Trial) "Forestry Carbon Credit Trading Contract Model" (Trial) "Forestry Carbon Credit Trading Commission Management Measures" (Trial) "Forestry Carbon Credit Trading Fund Settlement Measures" (Trial) "Forestry Carbon Credit Trading Dispute Resolution Measures" (Trial)

C *Innovate Project and Prioritizing Quality Control*

The CGCF has engaged in interesting project development, implementation, and management experiments, staying dedicated to voluntary reductions by enterprises, and has gradually established a unique public welfare project brand.

1 Mobilize Funds from Diverse Channels to Accelerate Project Implementation

Increasing carbon sequestration and reducing carbon emissions requires sooner realization of afforestation, forest management, and forest protection projects. The predominant difference between afforestation projects carried out by the CGCF and other ordinary afforestation projects is that the CGCF makes clear calculations of the carbon sink and production source so as to achieve real carbon sink increases, carbon emissions reductions. As of June 2012, the CGCF had received over 500 million yuan in donations from enterprises and individuals, which it used to establish more than 80,000 hectares of carbon sink forests in nearly 20 provinces (autonomous regions and municipalities) in China. Oil-yielding energy forests were planted in Yunnan and Sichuan, using Jatropha curcas as the main species and using Xanthoceras sorbifolia in Inner Mongolia. Moso bamboo forests with dual ecological and economic purposes were planted in Zhejiang Province. Forest management projects focused on increasing carbon sinks and reducing emissions were implemented in the Daxing'anling and Tangwang River of Yichun. Afforestation projects seeking to sequester carbon and yield additional ecological benefits were carried out in Gansu and Beijing. All afforestation and forest management projects were implemented in compliance with the developed technical standards including their accounting, monitoring, validation, and verification of carbon sinks, ensuring the quality of carbon sink accounting and monitoring throughout the entirety of the process.

2 Carbon Neutral Projects and a Professional Brand

The CGCF was assigned responsibility for the UNFCCC Tianjin Conference carbon neutral project in 2010. According to the calculations made by the Energy Economy and Environment Institute of Tsinghua University, there were an estimated total of 12,000 tons of carbon dioxide equivalent (CO_2–e) emitted during Tianjin Conference. The CGCF invested RMB 3.75 million yuan to establish 333 hectares of carbon sink forests in Xiangyuan, Xiyang, and Pingshun counties in Shanxi Province, which should be capable of absorbing the full carbon emissions mass of the Tianjin Conference over the next 10 years. The farmers in the project regions are expected to earn 2.6 million yuan in income for their labor contribution, and revenues from forestry byproducts and timber sales will equal 7 million yuan over the 20–year project management and operation period. The CGCF also organized a number of carbon neutralization projects for a series of organizations and for large-scale meetings of the International Network for Bamboo and Rattan (INBAR), China Green Companies Annual Conference, and the National Conference of General Directors of the Forestry Departments and Bureaus (see Table 10.2). The CGCF also established a host of individually donated carbon sequestration afforestation projects featuring different themes including the "State Counselor Forest for Carbon Sequestration," "Badaling Demonstration Forest for Carbon Sequestration," "China Building Materials Academy High School Carbon Sequestration and Science Promotion Forest," and "Model Worker Forest for Carbon Sequestration (Heilongjiang Xinxing Forest Farm)." The CGCF designed and sold a set of the world's first "Carbon Sequestration Gift Cards" during various holidays including the Spring Festival, Valentine's Day, and Adult Rites, etc., and set up nearly 40 individually donated carbon sequestration afforestation bases in cities and counties across the country. Anyone can log onto the CGCF website and browse through the selection of afforestation sites and species, and donations can be made by purchasing greeting cards or choosing tree planting directly by clicking on the link. A purchase certificate can be printed immediately. The system

TABLE 10.2 *Carbon Neutral Projects organized by the China Green Carbon Foundation*

No.	Project Title
1	Carbon Neutral Project for the 2010 UNFCCC Tianjin Conference
2	Carbon Neutral Project for the 3rd China International Forum of Ecological Civilization and Green Competition in 2010
3	Carbon Neutral Project for the National Conference of General Directors of the Forestry Departments and Bureaus in 2011
4	Carbon Neutral Project for the National Conference of General Directors of the Forestry Departments and Bureaus in 2012
5	Carbon Neutral Project for the National Autumn Winter Forest Fire Prevention Conference in 2011
6	Carbon Neutral Project for the 2011 China Green Companies Annual Conference
7	Carbon Neutral Project for the 2012 China Green Companies Annual Conference
8	Carbon Neutral Project for INBAR in 2010
9	Carbon Neutral Project for INBAR in 2011
10	Carbon Neutral Project for "Green Concert—Zero Carbon Concert Season" in 2010
11	Carbon Neutral Project for "Green Concert—Zero Carbon Concert Season" in 2011
12	Carbon Neutral Enterprise Project for Fujian Jianfeng Company in 2010
13	Carbon Neutral Project for the CGCF's Official Business Trips

achieves a modern level of professional accounting, regulated management, and transparency.

3 Launching the "Carbon Sequestration Action Plan in China"

On the 43rd World Earth Day in 2012, the CGCF launched the "Carbon Sequestration Action Plan in China." This plan was launched in cooperation with the Western Countries Forestry Company, the South -North Joint Forestry Property Rights Trading Company, Fulaisen Group Co. Ltd., and Zhejiang Couson Electronic Technology Co. Ltd., with a total donation of RMB 20.8 million yuan. The plan aims to spread the concept of green low-carbon actions, promote voluntary emissions reduction for enterprises, protect the natural environment, mitigate global warming, and protect the planet. Two programs have already been implemented, the "Nature Conservation Program" and the "Green Awareness Program."

The "Carbon Sequestration Action Plan in China—Nature Conservation Program" focuses on natural ecological conservation. The objective is to provide a better environment for the harmonious coexistence of humans and nature and to promote sustainable economic and social development through vegetation rehabilitation, forest protection, wildlife and habitat protection, nature conservation and scientific education, which enhances natural reserves and mobilizes society to actively participate in afforestation and biodiversity conservation.

The objective of "Carbon Sequestration Action Plan in China—Green Awareness Program" is to increase awareness of climate change, and specifically of carbon sequestration and green development, advocate for low-carbon production and low-carbon lifestyles, educate the public of ways to address climate change, establish carbon sequestration volunteer teams that will produce green carbon sequestration public cultural activities, and mobilize the public to contribute to climate change mitigation and adaptation supporting carbon offsets and carbon footprint elimination. The "Carbon Sequestration Volunteers Coalition" and the "Green Communication Center" were initiated by the Beijing Forestry University and remain affiliated with it, with more than several dozens of colleges, universities, and involved.

D *Carbon Credit Trading Pilot*

The CGCF has placed project implementation regulations and adopted a carbon credit production process that adheres to international standards and accounts for special circumstances in China (see Figure 10.3). The CGCF is diligent in ensuring that each ton of carbon credit can be traced back to its corresponding tract of forest, and that a one ton carbon credit seeks to incorporate other benefits including poverty alleviation, increasing farmers' income, biodiversity conservation, and ecosystem improvement. All carbon credits are properly accounted, monitored, registered, and certified, guaranteeing trading potential.

The CGCF has established a Carbon Credit Depository Platform based on preexisting trading standards and rules in conjunction with the Huadong Forestry Exchange to promote a national emissions reduction strategy based on forest carbon sequestration. On November 1, 2011, with approval from the State Forestry Administration, the Forestry Carbon Sequestration Trading Pilot Project was launched at the fourth International Forestry Expo in Yiwu City, Zhejiang Province. Relying on the trading platform provided by the Huadong Forestry Exchange, the CGCF provided 148,000 tons of carbon sinks (carbon

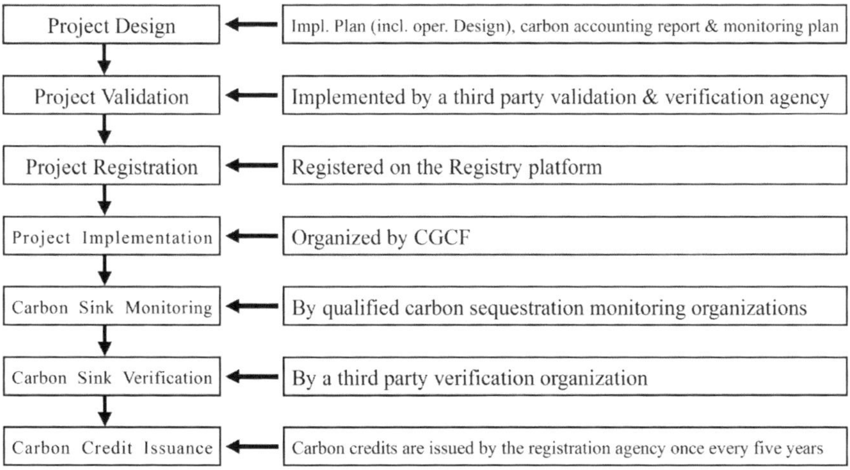

FIGURE 10.3 *The production process of the CGCF Carbon Credits for Carbon Sequestration Projects*

credits), which were all bought and sold on-site by ten enterprises including Alibaba, Geshan Construction Group Co. Ltd., Dezhengzhiyuan, Kaixuan Street, Hangzhou Qian & Wang Accounting Firm, Fuyang Timber Market, Longyou Foreign Trade Bamboo Shoots Manufactory Co. Ltd., Longyou Foreign Trade Bamboo Plant, Jiande Hongda Office Furniture, Zhejiang Mulaolao Toys, and Hangzhou Yuyue Investment Company. Carbon projects transaction validation was completed by Zhonglin Green Carbon Asset Management Center, supported by the Chinese Academy of Forestry. In accordance with international carbon trading rules, both trading project portfolios and validation statements have been published on the Huadong Forestry Exchange and the CGCF websites.

E *Cooperation with Multiple Parties for Scientific Research*

As China's first "Green Climate Fund" to address climate change, the CGCF has no previous examples to model and must innovate its own technology, policies and operations. In early 2011, CGCF established the nation's first Chinese Academy of Carbon Sequestration (the Academy), led by renowned domestic experts and scholars, which initiated research and international exchange on issues requiring urgent solutions. For example, the Academy has conducted studies "Improved Seed Breeding for Oil-Yielding Energy Forest Species," "Daqing Salix spp. Planting Patterns, Biomass Fuel Conversion, and Carbon Balance," "Low Carbon Afforestation Models for Eucalyptus," "Overseas Pilot Projects of Bamboo Forest Carbon Sequestration Afforestation Methodology,"

"Forest Management and Carbon Sequestration Project Methodology Development and Related Pilots by the Forestry Bureau of the Tangwang River of Yichun City." The CGCF also participated in specialized national forestry industry-specific studies for public welfare including a "Study on Technology for Increasing Forest Carbon Sink, Carbon Accounting, and Carbon Trade Market Mechanism" and a "Study on Carbon Transfer Accounting and Monitoring in International Trade of Forest Products and Forestry Carbon Sink Property Right." The CGCF has signed cooperative agreements with INBAR, the Nature Conservancy (TNC), Conservation International (CI), University of British Columbia (UBC), North Carolina State University, and many other international organizations and famous universities, and has carried out some independent training and cooperative scientific research activities. During 2011–2012, the CGCF organized a CI project named "Building Forest Carbon Sequestration Projects Capacity in China" and advanced its development of standards, staff training, and project implementation. The CGCF also established positive cooperative relations with research institutes and universities such as the Institute for Urban and Environmental Studies of the Chinese Academy of Social Sciences, Peking University, Chinese Academy of Forestry, Shanghai Jiaotong University, Beijing Forestry University, and Zhejiang Agriculture & Forestry University, among many others, which work collaboratively on joint forestry carbon sequestration studies and the Green Climate Fund.

F *Emphasizing Publicity and Carbon Sequestration Education*

The CGCF has employed an array of lively forums to explain forestry knowledge in addressing climate change to enterprises and individuals in order to get them involved in forestry welfare activities, to address climate change and to increase understanding of complicated concepts and professional jargon.

1. Organized media training courses targeting media professionals explaining carbon sequestration knowledge allow experts to accurately report on carbon sequestration and climate change news.
2. Carbon neutral concerts by the CGCF and the Beijing Municipal Bureau of Landscape and Forestry organized in 2010 and 2011 spanned a six months "Green Concert—Zero Carbon Concert Season" at the Forbidden City Concert Hall in Beijing. Famous artists were invited to participate and performance, forestry carbon sequestration information was printed on the back of all tickets, various posters were hung up at the Concert Hall, the emcee discussed forestry carbon sequestration, the audience

participated in a raffle, and carbon footprints were calculated on site, greatly increasing knowledge of the subject.

3. The CGCF and the Beijing International Studies University High School (National Ecological Civilization Education Base) jointly compiled and published the country's first curriculum on "Forestry Carbon Sequestration and Climate Change" to be taught in schools so as to publicize and explain the Green Climate Fund to high school students.

IV Outlook and Recommendations

In the face of the ever more sever climate threat and as a major GHG emitter, China established the GCF to effectively achieve strategic GHG reductions. The undertaking is historic and novel, full of both opportunities and challenges for China. Enacting carbon sequestration and emissions reduction through a public foundation requires further improvements to management and innovations of all activities to deal with the changing realities of climate change in China and around the world.

A *Strengthening Management and Standardizing Operation*

The CGCF is dedicated to "improving institutions, standardizing management, efficient operation, and remarkable achievements," and uses "fund raising through multiple channels and specialized implementation of projects" as its core strategy, focusing on execution of projects and credibility. The CGCF observes closely follows international climate change processes and serves the overall national climate change strategy. The CGCF is quick to adjust and change operations to better meet societies needs. By relying on a system of established technical standards and the registry platform for forestry carbon sequestration projects, the CGCF provides carbon sequestration accounting and monitoring technologies and project registration services for those enterprises, forest farms, and farmers who plan to carry out afforestation projects for carbon sequestration and for sustainable forest management. Some carbon sequestration projects in the voluntary carbon trading market will come from the CGCF Registry Platform's carbon sequestration and sustainable forest management projects, implemented by enterprises and farmers. The CGCF adheres to standardization, specialization, supervision and management of the entire process to ensure quality and effective projects. The CGCF will tighten financial management, welcome public supervision, and extensively, comprehensively, and clearly promote the CGCF to raise public awareness of climate change mitigation actions and environmental protection projects.

B *Expanding Fund-Raising Channels to Realize Sustainable Development*

For the CGCF to develop and expand effectively, it must raise funds through multiple channels, adopt various projects that address climate change and encourage enterprises to participate in voluntary emission reductions.

First, the CGCF must continue public fund-raising. Mobilizing a broader range of enterprises and individuals to actively make donations to the CGCF and implement afforestation activities for carbon sequestration and emissions reduction. Second, the CGCF must get national financial budget support. National financial contribution to the CGCF will be provided through procurement of services from public welfare organizations, which will increase carbon sequestration through afforestation, and expand the scope of support to other areas related to addressing climate change including green energy, education, and low-carbon economy trading activities. Third, the CGCF must receive financial contributions from high emissions industries. Mobilizing financial contributions from domestic and foreign high emissions enterprises so they can engage in voluntary emissions reductions in the form of capital contribution to CGCF for carbon offset is an essential financial building block for the Foundation. It also encourages more companies to voluntarily make their contributions to climate change mitigation and adaptation. Fourth, the CGCF must incorporate emissions reduction into the national carbon trading system. Proactively exploring the integration of green carbon sinks (credits) voluntary transactions with the national carbon emissions trading system will realize emissions reduction policies as actual operational carbon sinks and carbon source offsets. Such a model encourages enterprises to make effective use of low-cost green carbon credits to offset carbon emissions, and promotes the China's ecosystem services transfer to a market model, contributing to the realization of China's national voluntary greenhouse gas emissions reduction target, which provides a practical model and a set of experiences for the establishment and operation of the Durban GCF.

(This article was originally published in Chinese in 2012.)

CHAPTER 11

Weather Index Insurance and Commercial Applications in China

Su Buda, Tan Feng, Fang Yu, Thomas Fischer and Zhan Mingjin

Abstract

As disastrous extreme weather events triggered by climate change become more frequent each year, the economic losses caused by extreme weather events have led to ever higher socioeconomic costs. The application of adaptive tools and market-oriented applications are highly regarded by the international community as a response to the socioeconomic impacts of climate change and a way to strengthen natural disaster risk management. This article describes the advantages and disadvantages of a weather index insurance, comparing it with traditional property and casualty insurance. Various problems challenging development of weather index insurance products and implementation in the Chinese insurance market system are presented. A case study conducted by the China Meteorological Administration in Fujian Province offers the modus operandi and several developmental suggestions for a weather index insurance and its commercial application.

Keywords

Weather Index Insurance – Disaster Risk Management – Commercial Application – China

---

* Su Buda, National Climate Center of China Meteorological Administration, associate professor, research field: climate change impact assessment, climate risk management, meteorological index insurance. Tan Feng and Fang Yu, Nanjing University of Information Science and Technology, master degree candidate, research field: climate change and risk management. Thomas Fischer, National Climate Center of China Meteorological Administration, Guest Scientist, research field: climate change impact assessment. Zhan Mingjin, National Climate Center of China Meteorological Administration. Ph.D student, research field: climate change impact assessment.

 | DOI 10.1163/9789004274648_012

I Introduction

A meteorological disaster is a natural disaster that occurs frequently and has a particularly large impact range, involves high casualties, and causes large economic losses.

According to statistics from the Munich Re Group,[1] 88% of the major natural disasters recorded from 1980 to 2010 worldwide are primary or secondary meteorological disasters causing 59% of the total casualties. Meteorological disasters are the cause of 75% of economic losses and 91% of insured losses worldwide. Meteorological disasters have experiences a significant upward trend accompanying the increased frequency and intensity of extreme weather and climate events caused by climate change, wrecking greater havoc. In 2010, 828 meteorological disasters led to numerous casualties and serious economic losses, an increase of 160% compared with 1980 (317), a trend that causes further challenges to the sustainable development of the economy and society.

The rate of economic losses caused by climate change is increasing every year creating an ever greater need for international tools to reduce the effects of climate disasters. Since the first weather derivatives trading system was designed and applied to deal with agriculture disaster risk in the USA in 1997, developed countries in Europe have also tried to set up a weather index derivatives market, various meteorological catastrophe security plans, and an exchange market for meteorological disaster insurance products, meteorological index futures, and meteorological index options trading.[2] Agricultural weather index insurance has developed rapidly in some developing countries in Asia, Africa, and Latin America since 2002 with encouragement from the World Bank. Drought index insurance has emerged in India, Mexico, Malawi, Ethiopia, and Tanzania; flood index insurance products were introduced in Bangladesh and Vietnam; hurricane index insurance has appeared in the Caribbean Islands; and large-scale livestock catastrophe index insurance was created in Mongolia.[3]

China is located in the East Asian monsoon region, which has diverse natural conditions and a large variety of climate regimes, leaving it prone to various kinds of meteorological disasters. Due to the complicated procedures involved with common property damage insurances, traditional insurance products

1 Munich Re, Group Annual Report 2011. http://www.munichre.com/en/publications/default.aspx

2 Allianz Group and WWF, *Climate Change & the Financial Sector*. 2005.

3 The World Bank, *Index Insurance for Weather Risk in Low-Income Countries*, World Bank, 2007.

result in legal disputes leaving the product fairly inefficient and unable to except risk. The Chinese government struggles to offer financial support and lead reconstruction efforts after disasters due to shortages and funding delays. In 2007, China started to explore and develop ideas for weather index insurance. The Shanghai Anxin Agriculture Insurance Company Limited was first to develop a rainfall index insurance for watermelons during the rainy season. Many Chinese provinces have since begun to explore similar approaches for agricultural weather index insurance, and have made some achievements. In 2009 and 2010, the Ministry of Agriculture implemented rice planting weather index insurance and wheat planting weather index insurance pilot projects in Anhui Province, aiming to cover drought and flood disasters.[4] Although only a few insurance products were developed for the pilot projects, with only a limited area covered, they provided new and effective ideas and methods for agricultural insurance in China.

A *Weather Index Insurance's Strengths and Weaknesses*

Weather index insurance is based on the index of one or more meteorological conditions. The insurance is a contract through which a quantitative relationship between meteorological conditions and the extent of damage is developed and protected against. When the weather index reaches a certain level, the insured person or entity receives a standardized compensation. Meteorological indexes such as extreme temperatures, precipitation, wind speed, typhoon intensity, air humidity, among others, are generally used for weather index insurance.

Weather indexes, which are based on observation data from weather stations, are more objective and can be used for larger areas than the assessment of damage used for traditional property insurance. Weather index results are difficult for insured or insurer to change. The insured cannot change the meteorological condition. Weather index insurance is less likely to result in insurance fraud. Information asymmetry is also a smaller problem than in traditional insurance contracts, because the additional information that can be lost due to human screening (also known as adverse selection) can be avoided. Weather index insurance can also avoid large investments in manpower and material resources for large-scale disaster checking done for traditional property insurance, reducing operating costs and shortening the claims cycle. Weather index insurance contracts are standardized and transparent, insurance capital is

4 Chen Xiaomei, "Application Study of Index Based Weather Insurance in China," *Journal of Finance and Economics*, 2011 (9): 90–92.

allocated based on need. Weather index insurance transfers catastrophe risk onto the insurer and reduces the insured person's (e.g. the farmer's) risk.

Although weather index insurance ameliorates some of the drawbacks of traditional property insurance, it cannot fully cover the actual losses the insured person suffers from a disaster. The insurance company and insured person still face basic risks. For example, when a region is affected by a disaster, each insured person deals with their own vulnerability, their different capacity to withstand the disaster, and the different levels of damage suffered. Households insured with a weather index insurance are all compensated the same amounts when the weather index is triggered regardless of their differences. If one household is more severely affected than others, the compensation is nevertheless the same. Even if a household is not affected, it will still get the same compensation as soon as the weather index is triggered.

The following three issues clarify the reduced risks associated with weather index insurance. Weather indexes must improve the accuracy of their loss calculations. To ascertain the quality of the weather index, more and better maintained meteorological stations are needed. The weather index depends on the density distribution of meteorological stations. For example, temperature distribution is generally evenly distributed, but meteorological stations are not yet distributed densely enough to properly diminish uncertainties in the temperature index. The precipitation distribution is mostly unevenly distributed, meaning the areas covered by a precipitation index insurance product need more meteorological stations, allowing for a coefficient of weather index and the damage losses greater than 70%. The rate of loss, pricing relationships, and the basic risks when establishing an insurance product pricing model should be quantified. The insured party and the insurance company jointly accept the basic risks when entering an insurance contract. The insured party risks that actual loss may not be totally compensated, because when the weather index is triggered the losses are compensated in accordance with contract pricing.

1. A separate basic risk insurance contract should be designed in addition to the weather index insurance product. Regions with high basic risks and underdeveloped meteorological station infrastructure cover basic risks of the insured using additional contracts, like traditional property insurance from the government.

II Weather Index Insurance Market

The world's first weather derivatives trading started in the 1990s. Since then the meteorological disaster risk insurances market has developed rapidly.

Enterprises and individuals have increasingly demanded weather index insurance products. Many governments have turned their attention and support toward the development of these markets. The following elements should be considered in the design of weather index insurance products taking into account its new modern approach to disaster risk management. First, regions and industries that have weather index insurance products available should record historical materials about meteorological disaster impacts (such as the area damaged, number of affected people, and economic losses). Second, meteorological stations' current status (distribution and data availability) should be assessed in order to reduce basic risk, since the establishment of a weather index requires high quality observation data. The uniformity and availability of the data should be maintained at a certain standard. Third, factors causing regional meteorological disasters need to be identified, as each weather index insurance product only applies to losses caused by a specific type of disaster. When identifying the main factors of disasters, only the losses from a single factor or event should be focused on at a time. Fourth, the index scheme should be simple, scientific, and easy to understand for the insured person. The index should match actual disaster losses and confirm that the actual compensation is accurate and efficient.

Global meteorological disaster risk insurance is developed at the market floor and over-the-counter. Temperature index insurance trading makes up the overwhelming majority of transaction contracts and trading turnovers. Floor trading has increased significantly in recent years, while over-the-counter trading has been decreasing. This is mainly due to standard floor trading transactions and liquidation rules that reduce the transaction cost and the trading partner's credit risk. Over-the-counter trading contracts lack a fixed pattern making them relatively flexible and convenient, which is more suitable for the trial period of trading newly developed weather index derivatives that are not mature enough for the market. Over-the-counter trading is a good supplement to market floor transactions.

The frequency, magnitude, and economic losses of meteorological disasters in a certain region are the most important factors determining whether a scalable commercialization of the weather index insurance market is feasible. The main issue is to develop and design a market model fit for meteorological disaster risk insurance trading so that the systemic risk and the basic risk can be transferred and shared reasonably. A lack of recognized product pricing models is bottlenecking and blocking further development of the meteorological disaster risk insurance market. A pricing model cannot be directly applied to weather index insurance products because their market is different from a financial derivatives trading market in that the weather index has no monetary value. The weather index must be first monetized. Many different theoretical pricing models are used in the weather index insurance products market. Insurance pricing is generally based on two principal parameters, an estimate of damage probability, and an estimate of the risk for the insurance trading partners. Insurance companies do not often use complex pricing models, but derive a fair price based on historical data, and then add a risk coefficient. The risk coefficient should objectively reflect insurance companies' risk preference, as well as financing costs and the insurance product management and operation costs.

No meteorological disaster risk insurance market and trading products that cope with the effects of climate change have been developed in China. Development and design of weather index insurance products for the Chinese market must consider the following three aspects. First, risk of damage to individuals and businesses should be minimized. Second, reasonable market conditions should be established by promoting the completion of a trading market and by increasing the trading flows to reduce the cost of disaster risk compensation. Third, the systemic risk caused by extreme weather and climate events can be transferred and controlled effectively by engaging in pointed cooperative agreements between the government and the market, minimizing negative market impact.

III Weather Index Insurance Products in China

Linking the national disaster plan, risk management measures, and the weather index insurance market is critical for post-disaster reconstruction and timely compensation of disaster losses. The Chinese insurance market provides risk management consulting services for the government and enterprises. Weather index insurance implementation has a prominent advantage compared with traditional agricultural insurance. There are many theoretical weather index insurance plans being researched in China, such as a storm index for rice production in Zhejiang Province and a frost index for apple production in Shanxi Province.[5] But specific implementation schemes are still bundled with policy-oriented traditional agriculture insurance products, which central and local

5 Liu Yingning, He Wenli, Li Yanli et al., "A Study on the Risk Index Design of Agricultural Insurance on Apple Florescence Freezing Injury in Shaanxi Fruit Zone," *Chinese Journal of Agrometeorology*, 2010, 31 (1): 125–129.

Lou Weiping, Wu Lihong, Yao Yiping, "Design of Weather-Based Indemnity Indices for Paddy Rice Heavy Rain Damage Insurance," *Scientia Agricultura Sinica*, 2010, 43 (3): 632–639.

governments support with premium subsidies. The Chinese trial products are markedly different from typical commercial weather index insurance products.

The China Meteorological Administration has cooperated with the Insurance Regulatory Commission and several international organizations over the past few years to design demand-oriented commercial weather index insurance for small and medium enterprises as well as rural and urban households. The China Meteorological Administration conducted surveys to investigate demand and promote the purpose of a weather index insurance product in provinces which are seriously affected by meteorological disasters. Due to the gravity of the economic losses, some districts were selected for a pilot study for which weather indexes were developed and an insurance product which suits the local conditions was designed. The results of the pilot study on weather index insurance for tobacco production in Longyan City, Fujian Province are presented to give a practical example of the implementation of weather index insurance in China.

A *Assessment of Needs*

Tobacco is one of the most important economic crops in Fujian, planted on about 100,000 hectares and grossing annual sales of more than 1.5 billion RMB. Tobacco is sold using set prices determined by quality in China. Tobacco plants have a long growth cycle and are vulnerable to frost, drought, heavy rains, flooding, hail, typhoons, and other natural disasters. Chinese tobacco companies often receive some emergency aid after suffering losses from major disasters. The emergency aid for the farmers cannot cover the losses because a formal risk security system is lacking. The demand for an effective weather index insurance product has been increasing over the last few years.

To better understand the cultivation and production operations of tobacco, a socio-ecological survey was administered to investigate the tobacco planting system, planting acreage, planting costs, pricing systems, and the main meteorological disaster risks in 11 towns around Longyan City, Fujian Province, one of the most important tobacco growing regions in China. The analysis reveals that labor costs, the fertilizer purchase, and parching costs account for 85% of total tobacco cultivation costs. Tobacco seeding costs, irrigation costs, and equipment rental costs account for much of the rest. Up to 80% of the cultivation area can be affected by frost, heavy rain, flooding, hail, and other extreme

---

Cheng Xiaofeng, Huang Lu, "Experience of Drought Index Insurance in Malawi and its Inspiration for Development of Sugarcane Insurance in Guangxi," *Journal of Regional Financial Research*, 2010 (10): 53–56.

meteorological events. Frost and heavy rain are the most frequent and damaging events extreme meteorological events.

B *Preliminary Analysis of Meteorological Data*

The meteorological observation data, spatially standardized and sorted with temporal resolution, was obtained by the Meteorological Index Research Institute. A manual data collection system from meteorological stations was constructed and data from research departments were also used in the research. Over the past five years, the Fujian Climate Center installed 167 automatic observation stations to supplement the personal station network in Longyan City.

Most parts of Longyan City have hilly terrain and experience a subtropical monsoon climate. The region exhibits sharply varying altitude and exposure. Seven personal work stations and 128 automatic stations were chosen by a usability audit to provide the basic data for the meteorological research. The seven personal work stations are located in the seven county jurisdictions of the study area. Each station covers an area of about 2730 square kilometers and has an average resolution of about 50 km. The data records have a history longer than 50 years of daily collection and less than 5% of the data is missing making it suitable for time-series analysis. The meteorological parameters and their probability were analyzed. Automatic stations cannot be used for the time-series analysis of meteorological parameters because their observation period is too short. Because their spatial resolution is less than 12.5 km and they spread across all types of terrain and altitudes, they nevertheless play a considerable supplementary role in distinguishing microclimate errors (local weather conditions) caused by different terrain and exposure.

C *Analyzing Reasons for Disasters*

The tobacco production cycle is long, requiring about 10 months from winter until late summer. Seed planting starts at the end of November and early December in greenhouses. Transplanting onto fields is done in February and March. The tobacco is harvested and dried in July and August.

Because of local terrain and conditions, tobacco is mostly grown in low-lying areas where rainstorms or floods occur often. In Longyan City, rainstorms and floods are very frequent and are mostly concentrated during May and June. The tobacco plants grow high and almost ready to harvest during those months, but if they get damaged (bent or hurt), the quality decreases or in some cases the product cannot be recovered. Another threat is frost, which negatively affects tobacco plants in late February or March when it occurs. If the seedlings have already been transplanted onto the fields, one short frost can destroy the plant or at decrease its resistance to diseases. The tobacco production in Longyan City suffers its greatest damages and losses from rainstorms and floods, followed by frost and hail, and finally from pests and diseases.

D *Analysis of Disasters*

From 2002 to 2010, every tobacco farmer was affected by rainstorms, floods, frosts, and hail, in some way. Farmers lost an average of more than 500 kg of tobacco each, leading to economic losses of around 5,000–10,000 RMB. Tobacco production suffered from widespread frost, hail, and floods leading to heavy losses over the past few years.

A disaster evaluation index system was constructed to analyze the degrees of risk, exposure, and vulnerability in accordance with meteorological disaster risk parameters. By correlating and indexing the intensity, frequency, and location of heavy rains and frost disasters with tobacco production (per unit area) and its rate of increase or decrease, various statistical analysis methods established a disaster risk evaluation model, displayed in Equation 1. In this Equation, $Risk_i$ is the risk index for the region *i*, indicating the degree of tobacco production disaster risk in region *i*. H_i, E_i, and V_i are the risk, expose, and vulnerability indexes for region *i*. W_h, W_E, and W_v are the weighted values of the three indexes adopting an analytic ranking process.

Equation 1:

$$Risk_i = H_i W_h + E_i W_e + V_i W_v$$

E *Designing Weather Index Insurance Products*

The commercialization of a weather index insurance product remains a challenge after the index is designed. Using the data from personnel and automatic observation systems as well as information from the databases of extreme meteorological and climate events, the parameters and risk degrees are statistically calculated relating the developed index to disaster risk. The weather index is then used to define thresholds, which correspond to different levels of economic loss, and can be used to determine the insurance premium.

The rate of economic losses needs to be considered alongside the operational management and risk protection costs when determining the premium rate of weather index insurance. To calculate the premium rate, the probability of a disaster event, the quantitative estimates of the profitability of the insurance product, the company's operating expenses, as well as the risk safety factor must all be accounted for. In Equation 2, *probability* means the probability of the disaster event, *profit* is the insurance profit rate, *cost* is the operating costs, and *risk* stands for the risk safety factor.

Equation 2:

$$premium\_rate = \frac{probability}{profit + cost + risk}$$

F *Index Product Practice*

A low, medium, and high premium rate (e.g. 375 RMB/hectare, 1500 RMB/hectare, and 15,000 RMB/hectare) are developed with different thresholds and for the two weather indexes (rainstorm flood index, and frost index) to account for differing extents of meteorological disaster. Changting County in Longyan City suffered from a rainstorm in 2010. The rainstorm triggered the highest premium rate on the rainstorm flood index at 19 out of 22 automatic weather stations. According to the proposed premium rates, the insured tobacco farmers near these stations would get the highest compensation of 15,000 RMB/hectare. If the total tobacco production costs are 27,000 RMB/hectare, more than 55% of the theoretical economic losses would be compensated.

Considering these results, the combination of a national disaster plan and risk management measures with weather index insurance products would help to lower disaster risk and related economic loss potential. Researchers and insurance companies should further explore weather index derivatives as a significant means for post-disaster reconstruction and risk reduction.

(This article was originally published in Chinese in 2012.)

CHAPTER 12

Development of Renewable Energy and New Energy in China and Market Prospects

Wang Yu and Zhang Xiliang

Abstract

The development of renewable energy and new energy plays an important role in adjusting the energy structure, reducing greenhouse gas emissions, and protecting the environment in China. This paper offers an analysis of the development of renewable energy and new energy since the release of the *Law on Renewable Energy* in 2005, and an evaluation of hydropower, wind power, solar energy, nuclear energy, and biomass energy technological development in China. Finally, it provides a brief review of China's major renewable energy and new energy policies and their effects, as well as the development prospects for these energy production methods.

Keywords

renewable energy – new energy – greenhouse gas emissions reduction – cost

I Development of Renewable Energy and New Energy in China

China has advanced development of renewable energy and new energy since the signing of the *Law on Renewable Energy* in 2005. As of the end of 2010, annual consumption of renewable energy and new energy in China, which includes biogas, solar thermal energy, and non-commodity renewable energy, totaled around 300 million tons of coal equivalents, accounting for 9.6% of the total energy consumption in 2010.

* Wang Yu, Ph.D. at Research Institute of Energy, Environment and Economy, Tsinghua University. Research areas include analysis and assessment of renewable energy technologies and strategic planning of renewable energy.

Zhang Xiliang, Head of Research Institute of Energy, Environment and Economy, Tsinghua University, Ph.D. and Professor. Research areas include technology assessment of renewable energy, automobile energy strategy, energy prediction and early warning.

 | DOI 10.1163/9789004274648_013

Installed hydropower capacity reached 210 GW (gigawatts) as of the end of 2010, and annual generation capacity mounted to 650 billion kwh, which is the same as 208 million tons of coal equivalents, accounting for 6.3% of the total energy consumption. China's hydropower installed capacity and power generation have ranked No. 1 in the world for years. During the 11th Five-Year Plan period, new wind power installed capacity doubled for four consecutive years, and reached 18.9 GW in 2010. The cumulative installed capacity surpassed 40 GW, the world's second greatest installed capacity. China has 80 manufacturing enterprises producing all wind power equipment. The rate of local material use on 1.5 MW units has exceeded 70%. On-grid price of wind power is half the price of combustion power units. China's solar PV modules constructed over 8 GW in 2010, accounting for 53% of the global solar cell production. Installed solar capacity only reached 520 MW (megawatts) in 2010, and the cumulative installed capacity rested at 890 MW. The collective area used by solar water heaters grows at a rate of 30 million square meters annually, presently totaling approximately 160 million square meters. It can replace 30 million tons of coal equivalents derived from fossil energy, and has remained second to none in the world. China's biomass has gradually adopted diversified development. Biomass power generation capacity reached 5.5 GW in 2010. All small, medium, and large biogas projects in rural areas generated 13 billion cubic meters of gas, and provided clean, high-quality cooking fuel for more than 80 million rural residents. Annual ethanol production capacity reached 1.8 million tons, and biodiesel production surpassed 50 million tons. China installed 1.74 GW of new nuclear power facilities in 2010, and cumulative installed capacity reached 10.82 GW. The government also approved the construction of 34 nuclear power plant units, 28 of which have started construction. All together China's current and projected nuclear power accounts for 40% of the world's capacity under construction, making it No. 1 in the world (See Table 12.1.)

II Evaluation of Renewable Energy and New Energy Technological Developments

A *Hydropower*

Hydropower is a major provider of clean renewable energy. China has abundant water energy resources, theoretically 694 GW of reserves but feasibly 542 GW of developable resources, ranking first in the world. China's hydropower units have been designed and manufactured at the most advanced level in the world since the 1980s, and are moving toward international markets. China's

TABLE 12.1 *Current renewable energy in China*

	Application	2005	2010	Growth rate
Hydropower	Power generation	117 GW	210 GW	79.5%
Wind power	Power generation	1.26 GW	44.7 GW	3447.6%
Solar energy	Photovoltaic power generation	70,000 kilowatts	890,000 kilowatts	1171.4%
	Heat utilization	up to 80 million cubic meters of collector area of water heaters	160 million cubic meters of collector area of water heaters	100%
Biomass	Power generation	2 GW	5.5 GW	175%
	Fuel	1.02 million tons of ethanol	1.8 million tons of ethanol	76.5%
		50,000 tons of biodiesel	500,000 tons of biodiesel	900%
	Biogas	8 billion cubic meters	14 billion cubic meters	75%

Source: The State Electricity Regulatory Commission (2010), Renewables 2011 Global Status Report. www.ren21.net.

hydropower installed capacity exceeded 200 million kilowatts in 2010, after Xiaowan No. 4 was put into operation. China is not only a superpower with the world's largest installed hydropower capacity, but also a nation with the largest and fastest growing construction of projects. China has gradually become the global center of hydropower innovation.

Hydropower technology can be divided into large-scale and small-scale hydropower installations using 10–30 MW as a benchmark. Both are mature power generation technologies, and have been applied widely throughout the world. Small hydropower stations are usually installed in riverbeds. Because they minimally interfere with the river, they are environmentally friendly and they are mostly applied to supply power for rural residents. Large-scale hydropower projects are more complex, usually affecting the river downstream of the project. Hydropower technology has a better cost-competitive advantage and greater environmental benefits than conventional combustion power. First, hydropower can provide clean energy, easing the pressure of insufficient energy supply, and supporting economic and social development. Second, hydropower is an important means of promoting energy conservation and emissions reduction, furthering reducing environmental pollution. Third, hydropower resources are mainly distributed in western regions, bringing an unprecedented opportunity for hydropower development to these regions that is conducive to regional coordinated development. Average power generation cost for China's large-scale hydropower stations is about 0.15–0.20 RMB/kWh, while that of small-scale stations is about 0.21–0.24 RMB/kWh. Both are lower than the rate for combustion power, offering strong market competitiveness. But it is important to note that the on-grid price mechanism for small hydropower stations has not yet been established. Maintenance and management of the power stations remains weak. Ecological and environmental protection concerns have become increasingly prominent as well as concerns over immigration arising from development of large-scale hydropower stations.

B *Wind Power*

The Chinese government lends strong support to the development of wind power and has established a preliminary policy system covering the evaluation of wind energy resources, the industrialization of wind power equipment, on-grid price, and tax incentives. In 2009, the National Development and Reform Commission issued the *Notice to Improve the Policy of On-Grid Pricing of Electricity from Wind Power*, which established a system of regional benchmarks for wind power prices, eliminating vague pricing mechanism price variance. In 2010, China canceled a domestic wind power requirement that stipulated that more than 70% of equipment should be domestic, and encouraged a variety of funding sources, including foreign investment, to be injected into wind power construction. Thanks to these measures, China has witnessed rapid wind power development. The installed wind power capacity grew at a rate of 100% for 5 consecutive years from 2005 to 2009. As of 2010, the cumulative installed wind power capacity was greater than 40 GW in China, 10 years before the completion date for the 2020 wind power development targets set in the *Mid- and Long-Term Renewable Energy Plan*. China's enterprises have mastered the manufacturing technology of megawatt wind turbines, and are able to manufacture major parts domestically. Various 2–3 MW capacity unit prototypes have also been developed. Wind turbines cost around 5,000 RMB/kW; if 2000–2800 full load hours per year are assumed, the cost of wind power generation can be held at 0.5–0.6 RMB/kWh. Studies from the Global Wind Energy Council have shown that 60% of the decrease in wind power costs depends on large-scale development, and 40% on technological progress. By increasing turbine size the costs decline further promising wind power as the most economically competitive renewable energy technology in the future, except for hydropower. China has begun construction of eight 10 GW wind power bases, and led the first batch of offshore wind site bidding concessions in coastal regions. However, some bottlenecks have emerged such as intermittent resources, difficulties with large-scale grid combination, transmission, distribution, absorption, and power grid stability. Wind resource assessment is urgently needed to determine reserves for economic development, and to further improve assessment accuracy.[1,2]

C *Solar Energy*

China has developed a series of policies and measures supporting solar energy development. In 2009, the Ministry of Finance, Ministry of Science and Technology, and the National Energy Administration jointly issued the *Interim Management Measures for Financial Subsidies to Golden Sun Demonstration Projects,* which offer subsidies of up to 50% of the total investment for the project, as well as 70% of total investment in photovoltaic power generation systems in remote areas without electricity, and further support transmission and distribution projects. China's photovoltaic industry has witnessed rapid development along with the booming global PV market and policies offering preference to domestic production. China's solar cell module has produced 10 GW, accounting for 45% of world production, and installed capacity of photovoltaic power generation components has exceeded 800,000 kilowatts. The annual output of the photovoltaic cell manufacturing industry in China accounts for 40% of the global total. Solar power generation technologies include solar photovoltaic power generation and concentrated solar thermal power generation. China is still researching concentrated solar thermal and running experiments. The crystalline silicon technology for PV power is fully mature, while thin film technology is still in the demonstration phase. Monocrystalline and polycrystalline silicon technologies are two important ways to achieve mature crystalline silicon technology. The technical efficiency of Monocrystalline (about

1 Zhang, X.L., Chang, S.Y., Huo, M.L., Wang, R.S., 2009, "China's wind industry: policy lessons for domestic government interventions and international support," *Climate Policy* 9(5), 553–564.

2 Wang Zhongying, Ren Dongming, Gao Hu, *Report on the Development of Renewable Energy Industry in China 2008,* Chemical Industry Press, 2009.

15%, and with expected improvements up to 25–28%) is higher than that of polycrystalline, but the cost is correspondingly higher. Thin film technology can provide higher automatic control performance and resource production efficiency using fewer raw materials, and can be combined with buildings, but the power generation efficiency remains low. Using a life-cycle evaluation method, the payback period for photovoltaic power generation energy in different regions of China ranges form 2.8–5.1 years. At the end of 2008, the total cost of grid-connected photovoltaic systems was about 30 to 50 RMB/Wp, from which 66% is attributed to the cost of the photovoltaic cells themselves. The power generation cost of the grid-connected PV system was 1.5 RMB/kWh.[3] It is estimated that by 2015–2020 the cost of photovoltaic power generation system may shrink to 0.9 to 1.8 RMB/kWh. Research and development efforts of core technologies must be intensified, and more demonstration projects of solar power are needed to accumulate experience and improve the economical nature of the system.[4]

China has been an international leader in solar thermal utilization, and its solar water heaters that are at the core of the technology have reached advanced international levels. In the late 1990s, China's solar water heater industry experienced rapid development, and more than 100 major enterprises were engaged in the production of solar collectors and water heaters, producing over 40 million square meters of solar collectors with an output value of nearly 45 billion RMB. The Ministry of Finance and Ministry of Housing and Urban-Rural Development organized and implemented a renewable energy demonstration project, introducing the *Implementation Plan for Building Exemplary Cities with Renewable Energy Construction and Application*, and *Implementation Plan for Accelerating the Application of Renewable Energy Construction in Rural Areas*, promoting effective integration of production, design, and construction of solar thermal technology. Some studies show that solar water heaters can recoup the energy consumption for their production in just 1.1 years, the payback period for the initial investment of a solar air conditioning system is about 6 years, and that the cost of a solar energy collector system is in the range of 0.24–0.37 RMB. Solar thermal technology, compared

3 Wang Zhongying, Ren Dongming, Gao Hu, 2009, *Report on the Development of Renewable Energy Industry in China 2008*. Chemical Industry Press.

4 Wu Shu, 2009, "Evaluation for the Energy, Environment and Economic Effectiveness of Photovoltaic Industry," Master's thesis from Tsinghua University.

with traditional energy, has considerable cost advantages, and has a healthy future development potential.

D Biomass

Biomass power generation enjoys mature and large-scale development and modern technology, including direct combustion power generation, co-combustion power generation, landfill gas power generation, and waste incineration power generation technologies. The main technological features are shown in Table 12.2. At the end of 2010, the installed capacity of biomass power generation in China amounted to 5.5 GW. To regulate the price of biomass power generation and further promote the technology's development, the Chinese government released the *Notice on Improving the Policy on Power Generated from Agricultural and Forestry Biomass Prices*, which required that benchmark prices be implemented for on-grid electricity generated from agricultural and forestry biomass power generation projects. The rapidly developing biomass direct combustion power generation industry has imported its core technology from abroad. Biomass co-combustion power generation permits large-scale biomass use; however China has not offered clear preferential policies, so the use of co-combustion technology remains in the demonstration phase. Biomass gasification power generation systems have reached megawatts scales in China and landfill gas power generation demonstration projects have been completed. The cost of domestic direct combustion biomass power generation is 0.7–0.8 RMB/kWh. The cost of independently developed, small-scale biomass gasification power is 0.4–0.5 RMB/kWh. The cost of urban solid waste incineration power generation is 0.7–0.8 RMB/kWh. All are higher than the cost of combustion power. A variety of biomass power generation technologies face further challenges. Biomass direct combustion power generation technology is struggling with a slow localization process, high initial investment, high cost of collecting raw materials, and high project operation risks. Urban solid waste incineration power generation technology is also dealing with slow localization and high initial investment, as well as involvement by many stakeholders with diverse and unclear interests. Power generation from biogas in livestock and poultry farms and from industrial organic wastewater suffers from high on-grid price and a high generation cost. Biomass gasification power generation technology has achieved initial industrialization, but its system is not particularly efficient, the tar is not yet a common resource, and the single-scale internal combustion engine system must be large enough to be effective.

TABLE 12.2 *Efficiency comparison among typical biomass power generation technologies*

Type of transformation	Typical capacity	Net efficiency
Direct combustion power generation	10–100 MW	20%–40%
CHP[5]	0.1–1 MW	60%–90% All
	1–50 MW	80%–100% All
Hybrid combustion	5–100 MW (Present)	30%–40%
	>100 MW (New plants)	
Landfill Gas	<200 kW-2 MW	10%–15% Electricity
BIGCC[6]	5–10 MW Demonstration	35%
	30–200 MW In future	40%–50%

Source: OECD/IEA, 2008.[7]

In order to ensure energy security and reduce greenhouse gas emissions, the Chinese government has increased its financial support for the development of biofuels. Enterprises' ethanol production capacity has reached about 2 million tons/year. Ethanol gasoline (E10), which is mixed with ethanol, has been used in 10 provinces, and ethanol gasoline has reached 20% of national gasoline consumption. The annual biodiesel production capacity is over 500,000 tons, and China's ethanol use has reached 1.8 million tons. Liquid biofuel technology can be divided into three generations. The first generation of biofuel technology uses sugar, starch, oils, or animal fat as its raw material, producing ethanol or biodiesel through fermentation, transesterification, and other traditional conversion technologies. The second generation of biofuel technology uses cellulose such as straws or forestry waste as its main raw material. The third generation technology uses algae and other new materials as its raw material, and is still undergoing research and development.[8] Due to the large proportional cost of raw materials for first generation biofuels (60%-85%),

5 CHP, Combined Heat and Power.

6 BIGCC, Biomass Integrated Gasification Combined Cycle.

7 IEA, 2008, *Energy Technology Perspectives: Scenarios & Strategies to 2050* (Paris: IEA, 2008).

8 OECD/IEA, 2008, "From 1st to 2nd generation biofuel technologies: an overview of current Industry and RD&D activities," WWW.IEA.ORG.

they seriously impact the cost and development of first generation biofuels. As an alternative transportation fuel, the energy consumption and CO_2 emissions related to second generation biofuels are significantly lower than those of traditional gasoline and diesel using a full life-cycle analysis.[9] The International Energy Agency conducted a cost of biofuels estimate (Table 12.3), and found that using a relatively optimistic development curve, the production costs of cellulose ethanol and bio-synthetic diesel would significantly decrease after 2010, and would reach a relatively stable level in 2030, that is about USD 0.60/liter of gasoline equivalent. According to a relatively pessimistic development curve, the decline of production costs would be slow, staying around USD 0.65 to 0.70/liter of gasoline equivalent until 2050. The large-scale development of biofuel technology in China over the last 10–20 years has consistently faced high raw material costs, and a need for large breakthroughs in key second generation technologies.

TABLE 12.3 *Production costs of second generation biofuel technology*

Second-generation biofuel technology	Assumption	Production cost in Yuan/Ton[a]		
		2010	2020	2030
Cellulose ethanol[b] **(Biochemical methods)**	Optimistic	6806	5105	4934
	Pessimistic	6806	5955	5785
BTL[c]	Optimistic	10000	6500	6000
	Pessimistic	10000	8000	7800

Data source: IEA, 2008.[10]
a 1 U.S. dollar = 8 yuan (price in 2005);
b 1 liter of cellulose ethanol is 0.84 liters of gasoline equivalent;
c 1 liter of synthetic diesel is 1 liter of gasoline equivalent. BTL: biomass to liquids.

E *Nuclear Power Technology*

Nuclear power adopted an important role in future energy development because it is clean and economical. China will develop nuclear power as a

9 Ou, X.M., Zhang, X.L. Chang, S.Y., Guo, Q.F., 2009, "Energy consumption and GHG emissions of six biofuel pathways by LCA in (the) People's Republic of China," *Applied Energy* 86 (supplement 1), S197–S208.

10 IEA, 2008, *Energy Technology Perspectives: Scenarios & Strategies to 2050* (Paris: IEA, 2008).

large-scale alternative electricity generation option, and an important means of reducing greenhouse gas emissions. China had established ambitious development goals. Nuclear power technologies include nuclear fusion and nuclear fission. Fusion power is still undergoing research and development and predictions say it will not be ready for commercial application before 2050. The development of nuclear fission is generally divided into four generations, as shown in Table 12.4. Nearly all nuclear power plants that adopted first generation nuclear power technology have been shut down. Second generation nuclear power technology is currently operating in major nuclear power plants worldwide. Third and fourth generation nuclear power technologies are the most advanced and are the main technologies that will be implemented in future nuclear power developments.

China has mastered improved second generation nuclear power technology, the major technology currently used in the construction of nuclear power plants. China is still importing the AP1000, EPR and other third generation advanced nuclear power technologies from the United States and France. Once these advanced technologies have been fully integrated, new nuclear power plants and the utilization of uranium resources in China are expected to be much safer. As of the end of 2010, the installed capacity of nuclear power in China topped 10 GW, and 34 nuclear power plant units have been approved for construction, 28 of which have already started construction, accounting for 40% of all nuclear power plants under construction around the world.

Nuclear power has a higher initial investment (initial investment costs account for 50–60% of all costs, while fuel costs account for 20–30%), lower operating costs, and a longer design life than combustion power. The initial investment on second generation and third generation nuclear power plants is around USD 2,000/kW, with a payback period of 10–15 years. The cost of nuclear power generation is about 0.4 RMB/kWh. With the rising cost of fossil fuels for combustion power, nuclear power will gain an increasing competitive advantage. Third-generation nuclear power technology integration is not yet complete, the equipment manufacturing system is not yet fully developed, and other problems such as high initial investment and lack of nuclear development knowledge plague the field.

In early 2011, the Fukushima nuclear accident in Japan triggered a round of questions and discussions on nuclear safety in Japan and around the world. The question of balancing safety of nuclear facilities and meeting China's energy demand has been carefully considered in the nuclear power development process. China's nuclear safety supervision department and environmental protection department have established monitoring devices in operating nuclear power plants to ensure management safety. China's relevant departments have

TABLE 12.4 *Nuclear power technology development stages and characteristics*

	First generation	Second generation	Third generation	Fourth generation
Technological features	Early prototypes of reactor, almost all have been shut down	Major force in current commercial operation: LWR, CANDU, WER, etc.[11]	Advanced light water reactor: ABWR, System80+, AP600, AP1000, etc.	Major emerging nuclear technologies: supercritical water cooled reactor, ultra-high temperature gas cooled reactor, gas-cooled fast neutron reactor, sodium-cooled fast neutron reactor, molten salt breeder reactor
Development phases	From late 1950s to early 1960s	From 1960s to 1970s	Started from 1980s, and put to market at the end of 1990s	Just started, will be market-ready around 2035

said that they will learn from Japan when drafting nuclear power development strategies and plans, but China is determined to continue to develop nuclear power despite the Japanese nuclear accident.

11 LWR, light water reactor; CANDU, Canadian Deuterium Uranium Reactor; WER, water-water Energetic Reactor.

F *Summary*

The costs of new energy and renewable energy power generation technology is higher than that of combustion power with the exception of hydropower. But the costs of nuclear power and wind power are very close to those of combustion power. As these two technologies continue to improve and grow their economies of scale while fossil fuel prices rise, and as supportive national policies are passed, nuclear power and wind power will soon gain a competitive advantage over combustion power for some situations. Small-scale biomass gasification power generation technology enjoys low costs but due to raw material limitations, it is difficult foresee large-scale development. The cost of biomass direct combustion technology, waste incineration power generation technology and solar photovoltaic technology is more than twice the cost of combustion power, so cost reductions and efficiency improvements are essential.

III Policies for Renewable Energy and New Energy Development and Their Effects

Energy and environmental issues have become increasingly acute, making national energy strategy that actively develops new energy and renewable energy increasingly important. The Chinese government has always emphasized the development and utilization of new energy and renewable energy, and so has designed and developed a series of policies and measures to guarantee the healthy and rapid development of new energy and renewable energy, already achieving positive results.

A *Establish and Improve Laws, Regulations, and Systems*

The way to guarantee development of new energy and renewable energy, and to actively adjust the energy structure, is to improve laws and regulations. In 2005, the Standing Committee of the Tenth National People's Congress discussed and approved the *People's Republic of China Law on Renewable Energy*, which was implemented in January 2006. The law established five important systems addressing total targets, mandatory on-grid energy, tiered electricity pricing, cost-sharing, and special funding. With these five systems, China has built a policy framework to support renewable energy development. At the end of 2009, the National People's Congress added amendments to the law three years after its implementation, and clarified the requirements for national and local renewable energy development and utilization plans, stipulating

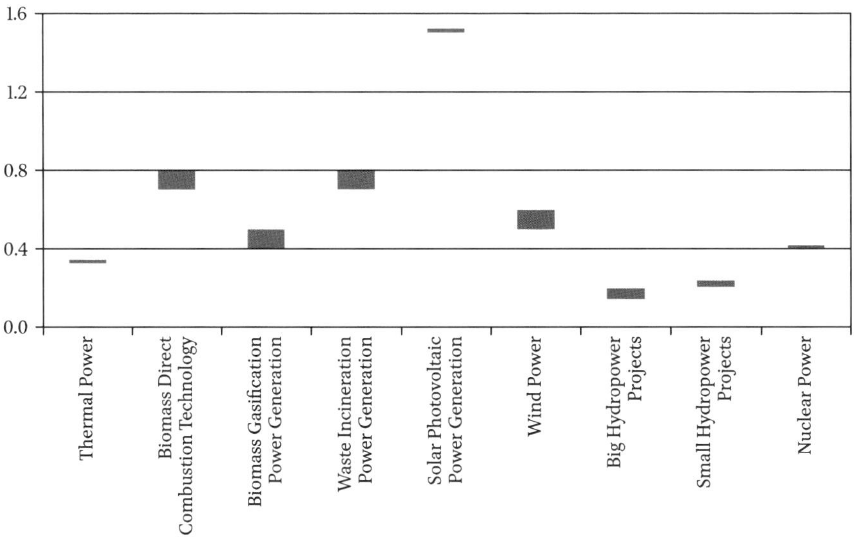

FIGURE 12.1 *New energy and renewable energy generation costs compared with combustion power*

the guaranteed acquisition system, whereby the government purchases all electricity generated through renewable energy technology, and explicitly stating that the government will establish a development fund for renewable energy.

B Strengthen Guidance with Planning

In accordance with the *Law on Renewable Energy*, China's National Development and Reform Commission has released a series of national energy development strategy plans focused on the development of non-fossil energy. The plans overall goal for new energy and renewable energy development is to "increase the proportion of total energy produced through renewable energy, provide electricity to remote and rural area populations currently without electricity, promote the energy oriented utilization of organic waste, and promote the industrial development of renewable energy technologies. The plans also clarified key development areas, conducted investment estimations and cost-benefit analyses, and proposed safeguard measures to ensure the plans' smooth implementation. At the end of 2009, the Chinese government adjusted the planning objectives to bring them in line with different technologies current situation.

C *Intensify Advanced Technology Development and Promotion Efforts*

China should research and master core technologies such as fast reactor design, nuclear fuel, and structural materials, seek advancements in sodium cycle research, and actively participate in the research and construction of the International Thermonuclear Experimental Reactor. Focus should be placed on low cost and large-scale development and utilization, such as large-scale wind power generation equipment, high performance-cost ratio solar photovoltaic cells and application technology, solar thermal power generation technology, technology for integration of solar power with buildings, as well as biomass and geothermal energy development and utilization technologies.

D *Economic Incentives*

Since 2000, China has offered a series of economic incentives to promote the rapid development of non-fossil energy. China published the *Notice to Improve the Policy of On-Grid Price of Wind Power Electricity*, setting four wind power benchmark prices at 0.51, 0.54, 0.58, and 0.61 RMB, respectively, thereby improving wind power price management. China issued the *Interim Measures for the Administration of Financial Assistance Funds for the Golden Sun Demonstration Project*, which offer assistance funds that cover 50% of the total investments on photovoltaic power generation projects and their supporting transmission and distribution projects, and 70% of the total investments on photovoltaic power generation projects in remote and isolated areas without electricity. China developed a cost sharing method renewable energy grid costs, clarified in detail the ways in which power generation enterprises take priority scheduling and purchasing generated renewable energy. The *Interim Management Measures for the Administration of the Special Fund for Developing Renewable Energy* stipulates that the special fund should be used for scientific and technological research, standard development, demonstration projects, energy in rural and pastoral areas, and independent power generation systems in remote areas and on islands. China also introduced a series of fiscal policies and tax benefit packages to promote the development of clean energy.

E *Increase Capital Investment*

During the 11th Five-Year Plan period, investment in power engineering construction shifted focus to non-fossil energy power generation in China. The proportion of investments in hydropower, nuclear power, wind power, from all power related investments increased from 29% in 2005 to 64% in 2010, and the investments in combustion power decreased from 227.1 billion RMB in 2005 to 131.1 billion RMB in 2010. China's investment in renewable energy was as high as 54 billion USD in 2010, more than the 41 billion USD invested in Germany, ranking first in the world.

With strong support from the *Law on Renewable Energy* and its supporting measures, China's new energy and renewable energy industry has been booming, and the development and utilization of various renewable energy technologies have undergone significant growth. The total amount of developed and installed new energy and renewable energy increased from 166 million tons of coal equivalents in 2005 to 260 million tons in 2009, making a significant contribution to bolstering China's energy security, and reducing greenhouse gas emissions. China's hydropower generating capacity in 2010 was 650 billion kWh, which is the same as the power generated by burning about 208 million tons of coal equivalents, thereby reducing carbon dioxide emissions by about 600 million tons. Installed solar water heaters took up 160 million square meters, replacing about 30 million tons of coal equivalents, eliminating about 8,000 tons of carbon dioxide emissions. China has constructed 13 new nuclear power generation units, raising total installed capacity to 10.8 GW, and generation capacity to 74.7 billion kWh, which is the same as the power generated by burning 25 million tons of coal, reducing carbon dioxide emissions by 73 million tons.

IV Future Prospects for Renewable Energy and New Energy Development

In the 12th Five-Year Plan period, the new energy industry is listed as a strategic emerging industry, placing focus on developing a new generation of nuclear power plants, using solar energy, photovoltaic power generation, wind power technology and equipment, developing a smart grid, and deploying biomass energy. The priority for new energy in the automobile industry is the development of plug-in hybrid vehicles, pure electric vehicles, and fuel cell vehicles. The project layout for new energy and renewable energy in the 12th Five-Year Plan foresees the following for China: Active development of hydropower.

A *Hydropower Technology Unquestionably Capable of Reducing CO_2 Emissions*

During the 12th Five-Year Plan period, China will accelerate development and implementation of large- and medium-sized hydropower projects, and actively develop small-sized hydropower projects, ensuring all the while protection of the ecological environment and sensitivity to social problems such as immigration and land acquisition. China will focus on the construction of large hydropower projects in the Southwest, and expand the scale of the "electricity from west to east" project. China will develop hydropower resources in small- and medium-sized rivers, and construct large hydropower stations in some key river basins including the Jinsha River, Yalong River, and Dadu River, in accordance with local conditions, providing a total hydropower capacity of 120 GW. Stronger grid supporting projects and effective wind power development. China will enhance the exploration and development of wind power resources, promote offshore applications of power generation models, solve wind power's problems linking to the power grid, encourage complementary wind and photovoltaic generation, and encourage a distributed power supply. In areas with rich wind resources such as eastern Mongolia and Xinjiang, China will construct six onshore, two coastal, and further offshore large-scale wind power bases. In areas with weaker wind resources, wind farms will be developed according to local conditions. The new installed capacity will be more than 70 GW.

B *Efficient Development of Safe Nuclear Power*

Although nuclear power technology is a CO_2 emissions reduction technology suffering from certain issues, it has the advantage of low emissions reduction costs. China will make use of both domestic and international markets to obtain uranium supply. By integrating third generation technology, China will improve large-scale manufacturing, installation, and construction capacity, accelerate nuclear power development in coastal provinces including Liaoning, Shandong, Jiangsu, Zhejiang, Fujian, Guangdong, and Guangxi, and steadily drive nuclear power generation in central region provinces such as Jiangxi, Hunan, Hubei, Anhui, and Jilin, boosting new construction of nuclear power generation to 40 GW.

C *Active Development of Solar, Biomass, Geothermal and Other New Energy Resources*

During the 12th Five-Year Plan period, China will further the popularization of solar water heaters, strengthen the research and development of solar photovoltaic technology, make efforts to develop its downstream technology market, conduct commercial demonstrations and promotions, and address the current solar photovoltaic production problem where neither end of the process occurs in China. In key provinces and autonomous regions such as Tibet, Inner Mongolia, Gansu, Ningxia, Qinghai, Xinjiang, and Yunnan, China will build more than 5 GW of solar power plants, promote biomass power generation in areas with rich biomass resources, encourage co-combustion of coal and biomass power generation, support power generation through urban solid waste incineration, push for biogas power generation in livestock and poultry farms, develop biogas power generation from industrial organic wastewater, and rationally distribute agricultural and forestry resource biomass power plants. China will also moderately develop geothermal energy in areas rich in geothermal resources.

It is expected that by 2020, the installed capacity of hydropower in China will reach 300 GW (including 75 GW from small hydropower projects), the installed capacity of wind power will reach 200 GW, solar power generation capacity will eclipse 20 GW, the total solar water heater collection area will reach 800 million square meters, total installed biomass energy capacity will reach 30 GW, annual biogas volume will reach 44 billion cubic meters, and annual ethanol and biodiesel production will reach 12 million tons. Including geothermal energy, ocean energy, and others into the equation, by 2020, the renewable energy used nationwide will be equivalent to 600 million tons of coal, and counting 70 GW of nuclear power, it is expected that the proportion of non-fossil energy out of primary energy consumption will exceed 15%. Realizing this objective will make a significant contribution to the structural adjustment of energy consumption in China, and play a crucial role in reducing greenhouse gas emissions and protecting the environment.

(This article was originally published in Chinese in 2012.)

CHAPTER 13

Development Potential for Wind and Solar Energy Resources in China

Zhu Rong

Abstract

This paper introduces the current state of wind and solar energy resource development and utilization in China. Existing challenges are introduced and power potential, spatial distribution, and development costs are analyzed. The results reveal that wind and solar energy resources' potential is enough to merit wind and solar power development. Finally, the paper discusses development and utilization safeguard policies for Chinese wind and solar energy, such as feed-in tariffs, cost allocation, smart grid development, distributed access, etc.

Keywords

wind energy resource – solar energy resource – grid connection – distributed development – electricity price – cost

I Current Wind and Solar Energy Development and Utilization in China

Countries around the world have adopted development and implementation of renewable energy as a significant measure for reducing fossil energy consumption, addressing climate change, conserving energy, and reducing emissions. In 2011, global renewable energy sources provided almost 50% of power capacity increase, and of that wind energy and photovoltaic technology contributed 40% and 30% of power generated by renewable energy

* Zhu Rong, a senior scientist at the CMA Wind and Solar Energy Resources Center, engages in numerical simulation and forecasting of wind and solar energy resources, atmospheric environment impact assessment, etc.

 | DOI 10.1163/9789004274648_014

respectively, and hydropower contributed 25%.[1] By the end of 2011, the total installed renewable energy capacity around the world exceeded 1,360 GW, which represents an 8% increase in one year. From the end of 2006 to 2011, power generating photovoltaic technology experienced the fastest development compared with all other renewable energy sources, increasing average installed capacity by 58% annually. Concentrated solar energy increased by 37%, and wind power by 26%. Some developed countries set specific renewable energy installation goals. The EU set a target of reaching 20% of energy consumption from renewable energy by 2020. Spanish plans foresee renewable energy consumption accounting for 22.7% of Spain's total energy consumption from 2011 to 2020, and clean energy sources providing 42.3% of its total power generation, placing it slightly ahead of the EU goal. Denmark has clearly stated that it will completely cut its dependence on fossil fuels for energy by 2050. The following 20 to 50 years will be ripe with important energy transformations.

Over the last ten years, along with rapid national economic development, energy consumption in China has increased every year (Figure 13.1). According to data from the BP Statistical Review of World Energy 2012,[2] primary energy consumption in China reached 2.75 billion tons of oil equivalent in 2011, accounting for 22.4% of total global primary energy consumption. Renewable energy consumption in China equaled 180 million tons of oil in 2011, accounting for 17.9% of the total global renewable energy consumption. Since the "Renewable Energy Law of the People's Republic of China" was issued in February 2005, the development and utilization of renewable energy resources has grown rapidly. By the end of 2011, China's total installed renewable energy resource capacity ranked first in the world at 282 GW, a quarter of which (70 GW) was provided by non-hydroelectric renewable energy. Of the 90 GW of new energy installed in 2011, renewable energy contributed one-third, and non-hydroelectric renewable energy contributed one-fifth. The 12th Five Year Plan set the following development goals for China's renewable energy for 2015: wind power generation shall reach 100 GW, providing an annual power generation of 190 billion kWh, and including 5 GW of offshore wind energy. Solar power generation shall reach 15 GW, with an annual power generation of 20 billion kWh. Adding the contributions from biomass, solar thermal, and nuclear power generation, the total non-fossil energy development shall reach 480 million tons of standard coal equivalent.

---

1 Renewable Energy Policy Network for the 21st Century, Renewables 2012, Global Status Report, www.ren21.net.

2 BP Statistical Review of World Energy June 2012, bp.com/statisticalreview.

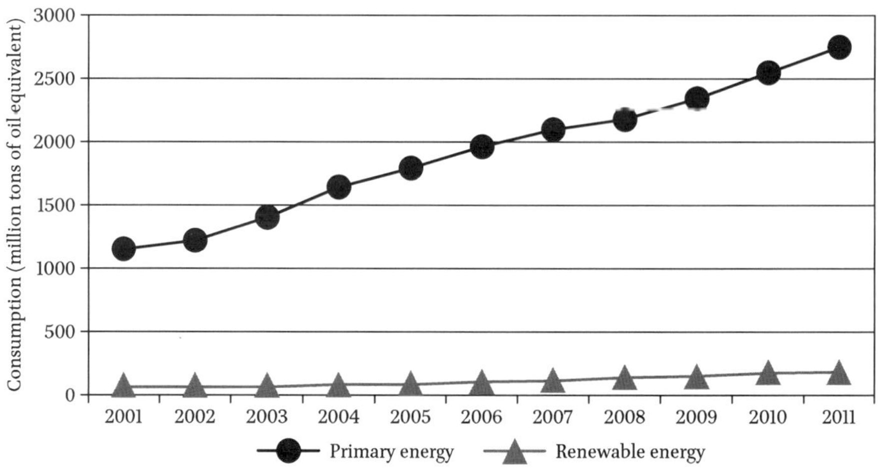

FIGURE 13.1 *Consumption of primary energy and renewable energy in China from 2001 to 2011* (DATA SOURCE: BP STATISTICAL REVIEW OF WORLD ENERGY 2012)

A *Current Wind Energy Resource Development*

From 2001 to 2011, the installed capacity of China's wind turbines increased from 0.38 to 62.36 GW (excluding Taiwan), due to the installation of 45,894 wind turbines, granting China the greatest installed capacity of wind turbines in the world (see Figure 13.2). China also ranks first, followed by the USA, Germany, and Spain, in the capacity of newly installed wind turbines (Figure 13.3). The accumulated installed capacity of wind turbines in the Inner Mongolia Autonomous Region reached 17.59 GW in 2011, accounting for 28% of the total installed capacity in China, placing it far ahead of other provinces. The accumulated installed capacity in Inner Mongolia, Hebei, and Liaoning accounted for 45.3% of that of the whole of that of the whole of the country capacity (Figure 13.4).[3] The development of China's wind power market focuses on and will continue to focus on large-scale onshore wind farms, on actively promoting the construction of offshore wind farm demonstration projects, and on discussing the development of distributed gird-connected wind power projects.[4] In 2010, Phase I of the first wind energy base in China with a power capacity higher than 10 GW was completed in Jiuquan, Gansu Province, after installation of 5.36 GW. Seven wind farms with power capacity higher than

3 Statistical data by Chinese Wind Energy Association.

4 China Wind Energy Development Roadmap 2050 issued by the Energy Research Institute of the National Development and Reform Commission and IEA.

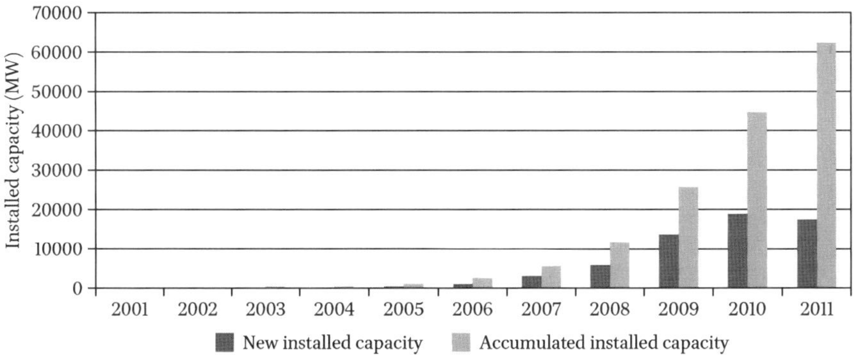

FIGURE 13.2 *Changes of new and accumulated installed capacity of wind turbines in China from 2001 to 2011*
(DATA SOURCE: STATISTICS OF INSTALLED CAPACITY OF WIND TURBINES IN CHINA 2011 BY CHINESE WIND ENERGY ASSOCIATION)

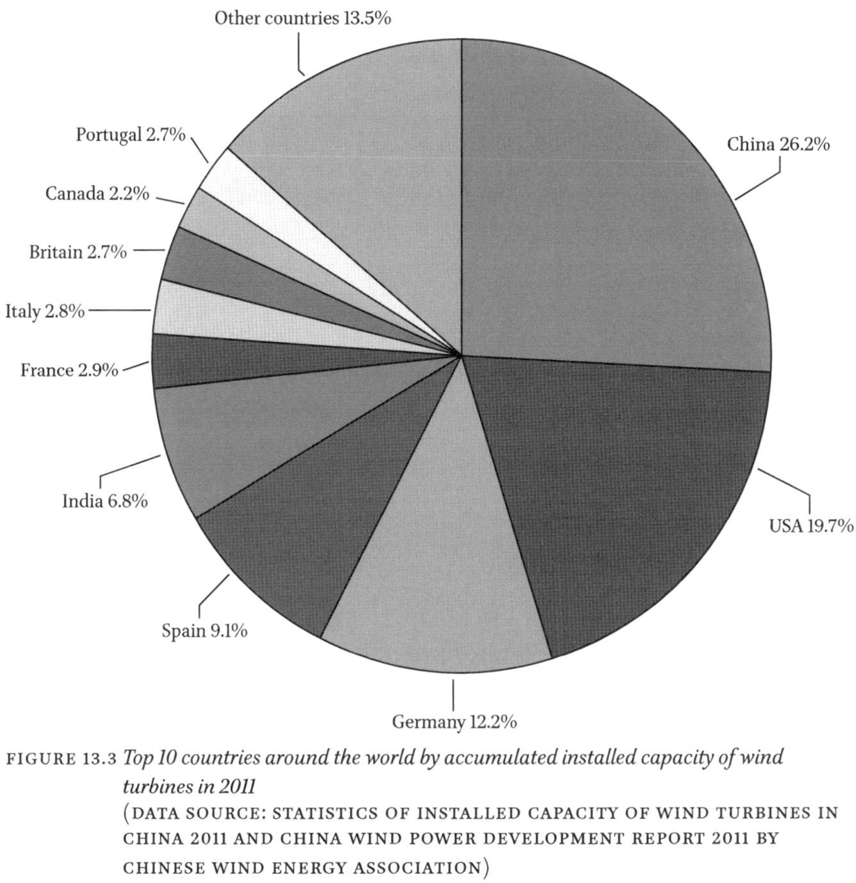

FIGURE 13.3 *Top 10 countries around the world by accumulated installed capacity of wind turbines in 2011*
(DATA SOURCE: STATISTICS OF INSTALLED CAPACITY OF WIND TURBINES IN CHINA 2011 AND CHINA WIND POWER DEVELOPMENT REPORT 2011 BY CHINESE WIND ENERGY ASSOCIATION)

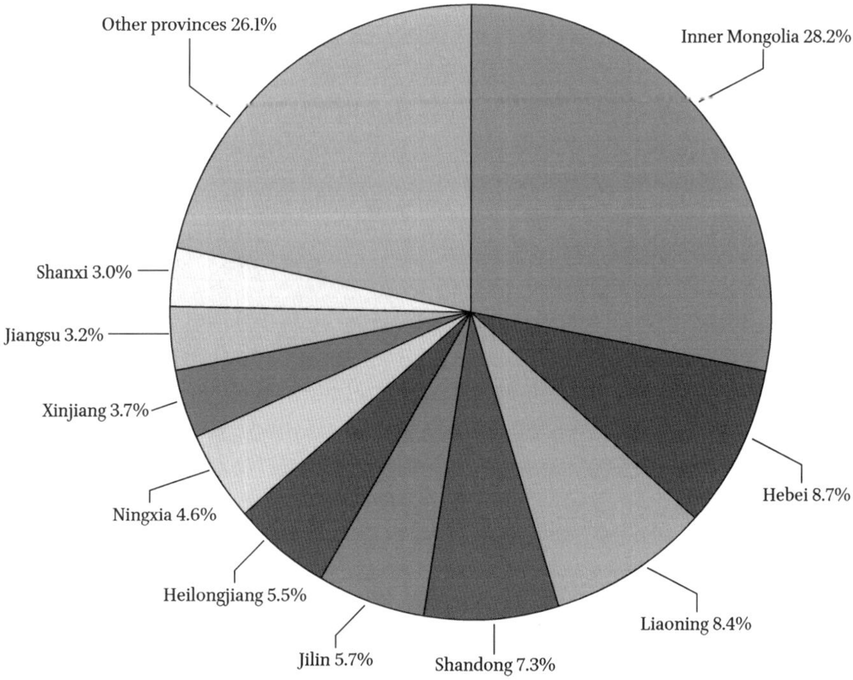

FIGURE 13.4 *Top 10 Chinese provinces by accumulated installed capacity of wind turbines in 2011* (DATA SOURCE: STATISTICS OF INSTALLED CAPACITY OF WIND TURBINES IN CHINA 2011 AND CHINA WIND POWER DEVELOPMENT REPORT 2011 BY CHINESE WIND ENERGY ASSOCIATION)

10 GW will be built in eastern Inner Mongolia, western Inner Mongolia, Bashang in Hebei province, Hami in Xingjiang, western Jilin, Jiangsu, and Shandong. The Shanghai East Ocean Bridge offshore wind power project, which is the first offshore wind farm in China with a power capacity of 100 MW, was officially completed and put into operation in 2010 as well. The completion of public bidding for the concessions for a 1 GW offshore wind power project signals the official launch of offshore wind power development in China.

Integrating and accommodating wind power within the general power grid China has gradually become the primary restriction on wind power development. Wind power development is highly concentrated in northeastern, northern, and northwestern China (the "three-norths"). This concentration causes a development and grid construction mismatch, a low local grid load, a poor power supply that cannot be flexibly adjusted, and suffers from an immature inter-provincial and inter-regional market. Grid integration and market accommodation of wind power in the three-norths has already become a prominent issue, and the unused wind energy resources lie in wait. The National Energy Administration has called for greater care in the operation and management of wind farms, and for active expansion of the wind power markets to improve the wind energy's utilization ratio, using measures such as power prediction and forecasting, improved wind turbine technology, better demand management, and wind power heat and energy storage.

B Current Solar Energy Development and Utilization

Solar energy is inexhaustible, clean, and safe, making it an ideal renewable energy resource. Due to severe power shortages in 60% of all counties with any

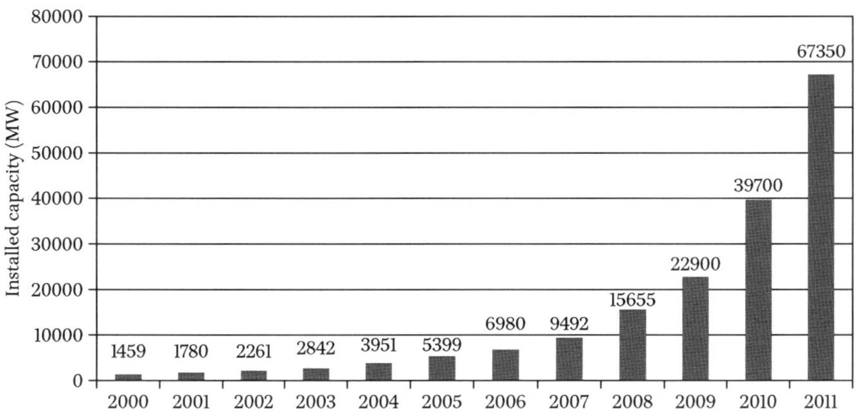

FIGURE 13.5 *Global accumulated installed capacity of PV from 2000 to 2011* (DATA SOURCE: EPIA MARKET REPORT 2011)

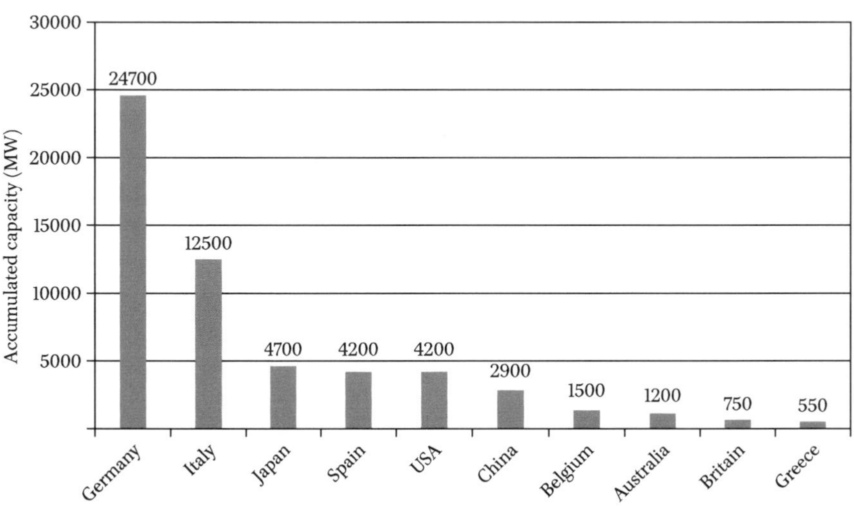

FIGURE 13.6 *Top 10 countries by accumulated PV installed capacity in 2011* (DATA SOURCE: EPIA MARKET REPORT 2011)

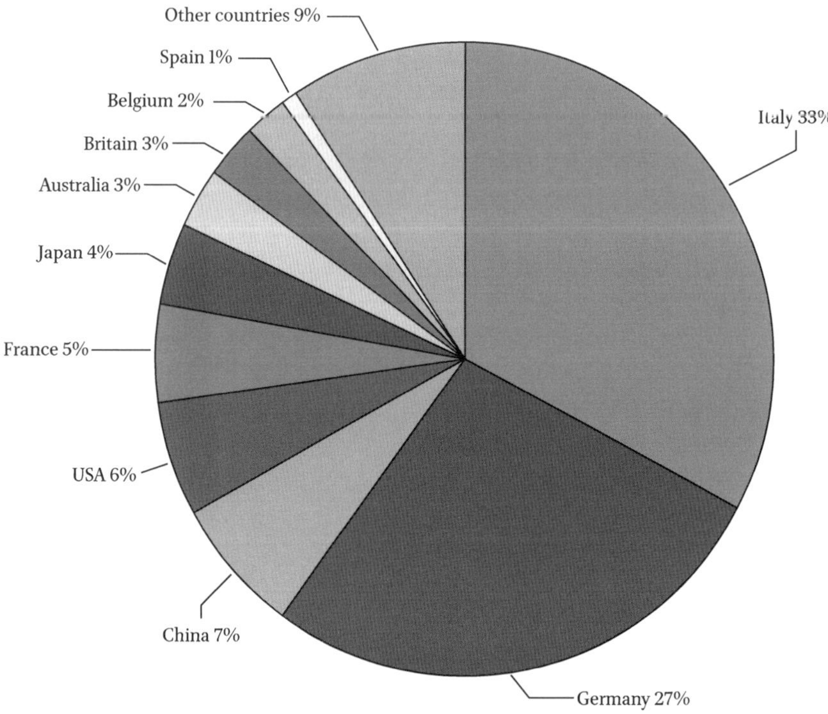

FIGURE 13.7 *Top 10 countries by on-grid and new installed capacity of PV in 2011* (DATA SOURCE: EPIA MARKET REPORT 2011)

access to power in China, the photovoltaic market has a great potential. The newly installed capacity of global photovoltaic has increased continuously and rapidly since 2000 (Figure 13.5).[5] Global on-grid photovoltaic power generation in 2011 was nearly 38 times what it was in 2000, up to 27.7 GW, 21 GW of which were generated in European countries. The countries with the greatest accumulated installed PV capacity rank as follows: first, Germany with 24.7 GW; second, Italy with 12.5 GW; fifth, USA with 4.2 GW; and sixth, China with 2.9 GW (Figure 13.6). The countries with the greatest new installation of on-grid PV capacity in 2010 rank as follows: first, Italy with 9 GW (or 33%); second, Germany with 7.5 GW (or 27%); third Chinawith 2 GW (or 7%); and fourth, USA with 1.6 GW (Figure 13.7).

After years of effort, a relatively complete industrial solar photovoltaic manufacturing system has been established in China, but PV technology and the

5 European Photovoltaic Industry Association (EPIA) Market Report 2011, http://www.epia.org/fileadmin/EPIA_docs/public/EPIA-market-report-2011.pdf.

product market are still mainly dependent on foreign countries. Due to the influence of the global financial crisis and the rise of trade protectionism internationally, photovoltaic products reliant on foreign markets has struggled.

Photovoltaic power generation maybe rapid at present, but its future may be restricted by the grid. Concentrated solar thermal power, which has the advantage of energy storage, can also be used as an alternative to traditional fossil energy, such as combustion power. By the end of 2011, the capacity of global concentrated solar thermal power plants under operation was over 1,700 MW. Concentrated solar thermal power generation remains in the early stages in China, and at present no concentrated solar thermal power plants are operating. The Eerduosi solar parabolic trough power plant, which has a capacity of 50 MW, is the first solar thermal power plant project in China, and has already gone through concession bidding. The project uses trough concentration technology for solar thermal power generation, with a planned total investment of RMB 1.6 billion and annual power generation of 120 million kWh. The 12th Five Year Plan of National Energy Science and Technology presented in December, 2011 proposes scaling solar thermal power generation technology so it can bear a foundational load, and suggests the following economic and technical targets for 2015: achieve a generating efficiency of no less than 15% for updated plants, lower generating cost to no more than 1.5 RMB/ kWh, the plant should be capable of 12 hour of continuous power generation per day, and rate of change of the power load during the generation period should be less than 5%. The plan further set a of establishing a large-scale solar thermal power generation demonstration project in China.

II Wind and Solar Energy Resources Distribution and Development Potential in China

Resource potential is measured as theoretical potential, technical potential, and economic potential. The total wind and solar resource amount refers to the quantity of theoretical energy calculated directly using the distribution of wind speed and solar radiation within a certain area. Technical energy refers to the energy obtained after deducting non-developable areas (such as topographic slopes, bodies of water, urban and ecological preservation zones, etc.) from the calculated area, using the theoretical total amount as the starting figure. Economic development quantity refers to the total amount of developable resources that are available under current technical and economic conditions and policies.

A Wind Energy Distribution and Development Potential

The China Meteorological Administration (CMA) built the national wind energy resource professional observation network from 2007 to 2011, which covers 31 provinces, cities, and autonomous regions, using 400 anemometer towers. The CMA used this data along with historical meteorological data to create numerical simulation and GIS spatial analysis technology, which it used for a new round of detailed assessments of wind energy resources in China. The latest results suggest that northeast of China, Inner Mongolia, the northern part of northern China, Jiuquan in Gansu province, and northern Xinjiang are the areas with the most abundant onshore wind energy resources. The Qinghai-Tibetan Plateau, the Yunnan-Guizhou Plateau, and southeastern coastal areas were identified as having good wind energy resources. The Sichuan Basin on the eastern side of the Qinghai-Tibetan Plateau, the Xinjiang Tarim Basin and the Dzungaria Basin, southern Nyingchi in Tibet, southwestern Yunnan, and Fujian's inland areas were found to have only limited wind energy resources (Figure 13.8). The technical potential of onshore wind energy resources (where the annual mean wind power density ≥300 w/ m^2) at a height of 50 m, 70 m, and 100 m throughout country is 2000 GW, 2600 GW, and 3400 GW respectively. Using 70 m as an example, the Inner Mongolia Autonomous Region has the greatest wind energy technical potential (about 1500 GW), and Xinjiang and Gansu are next (400 GW and 24 GW respectively). The technical potential is also great in onshore areas such as Heilongjiang, Jilin, and Hebei, and in coastal areas in provinces such as Hebei, Shandong, Jiangsu, and Fujian, which are suitable for large-scale wind farm construction. However, developable wind energy resources in other inland provinces lie primarily over mountain ridges or plateaus, which can only incorportae separate development.

The numerical simulation results for offshore wind energy resources (Figure 13.9) suggest that the Taiwan Strait has the most abundant wind energy potential, followed by eastern Guangdong, Zhejiang's offshore areas, and the north central Bohai Gulf. The areas with relatively little offshore wind energy resources include the Beibu Gulf and northwestern, southern, and southeastern offshore sea areas around Hainan Island. Although the Taiwan Strait has the most abundant offshore wind energy resources, it is difficult to develop wind energy there because of the area's small size and water depth from 5 to 50 m. Jiangsu province has a larger offshore area with water depth varying from 5 to 50 m. Although the wind energy resources there are weaker than they are in other provinces' offshore areas , the resource strength nevertheless meets the wind power development standard. The Bohai Gulf has the optimum mix of conditions for offshore wind energy resource development, because over half of the gulf area has a water depth varying from 5–25 m, and the water

TABLE 13.1 *Theoretical potential and technical potential for onshore and offshore wind energy resources in China*

Area	Theoretical Potential (GW)	Technical Potential (GW)
Onshore areas in China (excluding Qinghai-Tibet Plateau, at the height of 70 m)	5900	2600
Offshore areas in China (with water depth from 5–50 m, at the height of 100 m)	2500	500

Data source: Provided by CMA Wind and Solar Energy Resources Center.

never gets as deep as 50 m. The wind energy resources in Bohai Gulf are also very rich, as the mean annual wind speed is about 8.0 to 8.5 m/s and the mean annual wind power density is about 650–700 w/m^2. Based on the assumption that a 5 MW/km^2 installed capacity is possible where wind density reaches 300 w/m^2 and the analysis results of the Energy Research Institute of the National Energy Administration, it is believed that only 20% of all offshore areas within 50 km from the coastline can be used for wind energy resource development in China. The technological development potential for offshore wind energy resources in China is about 512 GW for areas with water depth varying from 5–50 m.

China has abundant wind energy resources. The theoretical potential of onshore wind energy resources is 5900 GW at a height of 70 m, and the technical potential is 2600 GW, excluding the Qinghai-Tibetan plateau. The theoretical potential of offshore wind energy resources is 2500 GW for areas with water depth varying from 5 to 50 m and at 100 m above the sea surface, and the technical potential is 500 GW (Table 13.1). Although the analysis and calculations of nationwide wind energy resource's technical potential is not yet complete, based on the theoretical potential and the available land and sea area, it is believed that China's wind energy resources are sufficient to support an installed capacity of over 1,000 GW using current wind power technology. Wind energy resources will not be the restricting factor in the development of China wind power, which has the potential to be an important component in the future energy and power structure.

The cost of wind power is primarily determined by wind energy resources, wind turbine and technology costs, operational and management technology, and power conversion costs. According to analysis in the China Wind Energy Development Roadmap for 2050 published by the Energy Research Institute

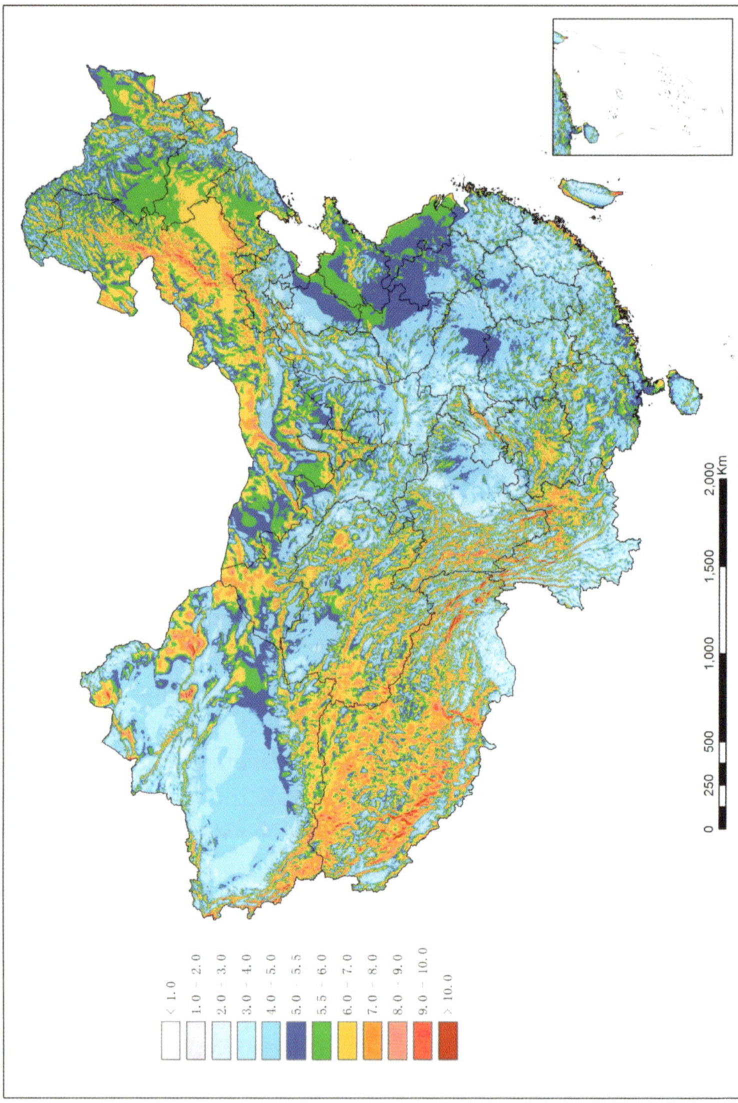
< 1.0
1.0 – 2.0
2.0 – 3.0
3.0 – 4.0
4.0 – 5.0
5.0 – 5.5
5.5 – 6.0
6.0 – 7.0
7.0 – 8.0
8.0 – 9.0
9.0 – 10.0
> 10.0
0
250
500
1,000
1,500
2,000 Km

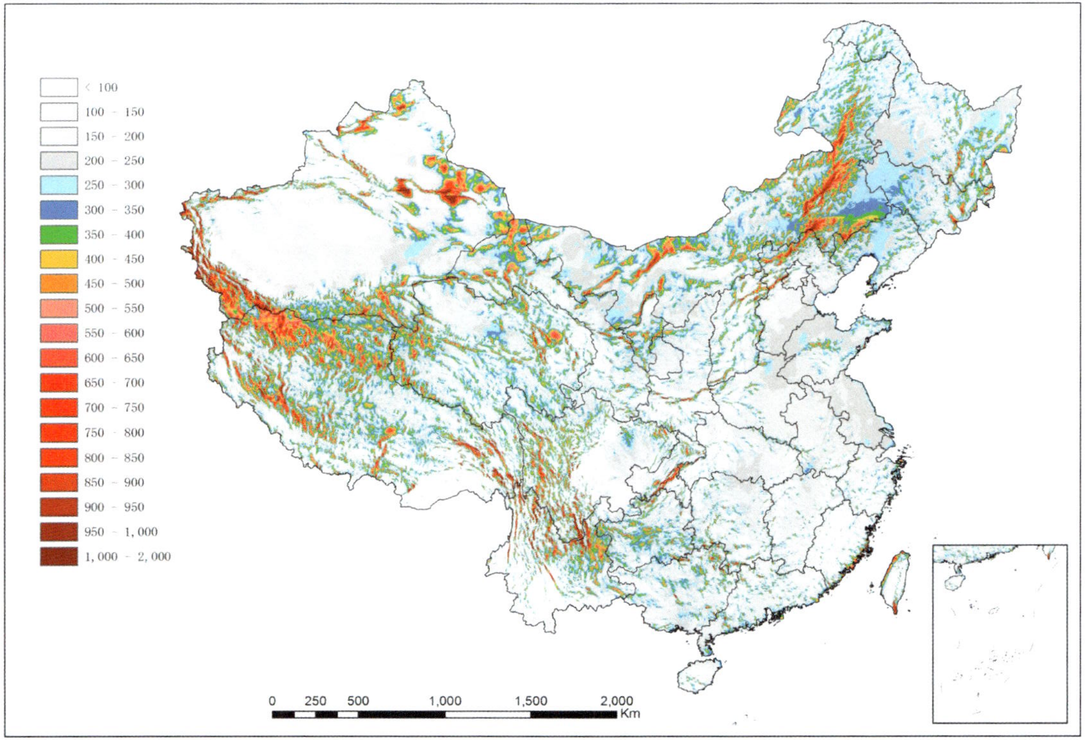

FIGURE 13.8 *Distribution of annual mean wind speed (left) and wind power density (right) in inland China with horizontal resolution of 1 km × 1 km and at a height of 70 m*
(PROVIDED BY CMA WIND AND SOLAR ENERGY RESOURCES CENTER)

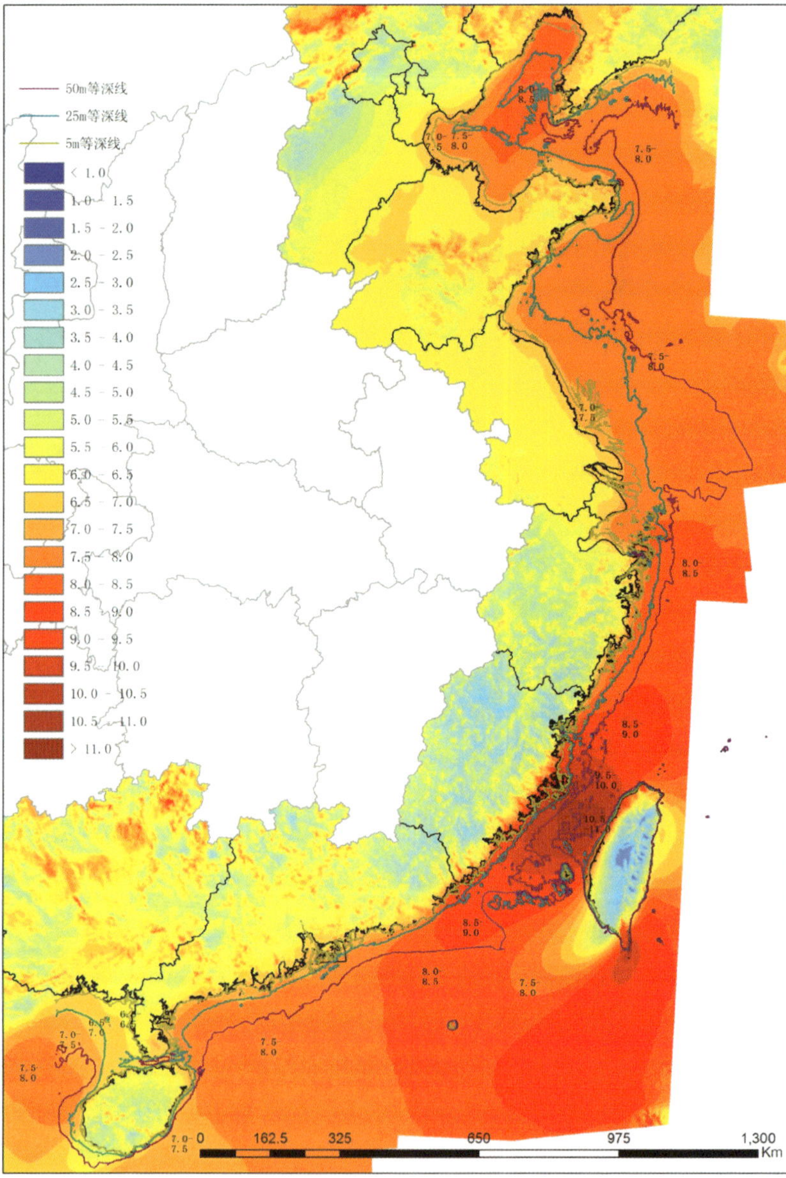
50m等深线
25m等深线
5m等深线
< 1.0
1.0 - 1.5
1.5 - 2.0
2.0 - 2.5
2.5 - 3.0
3.0 - 3.5
3.5 - 4.0
4.0 - 4.5
4.5 - 5.0
5.0 - 5.5
5.5 - 6.0
6.0 - 6.5
6.5 - 7.0
7.0 - 7.5
7.5 - 8.0
8.0 - 8.5
8.5 - 9.0
9.0 - 9.5
9.5 - 10.0
10.0 - 10.5
10.5 - 11.0
> 11.0
0
162.5
325
650
975
1,300
Km

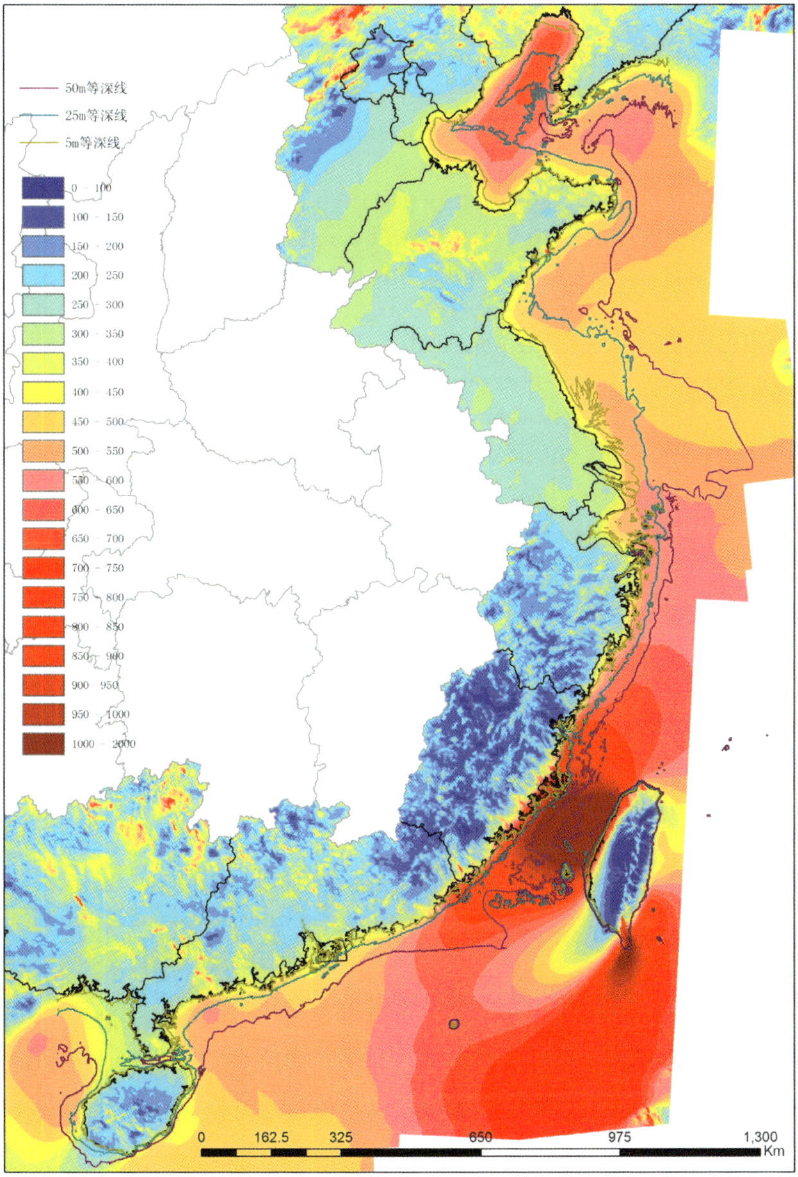

FIGURE 13.9 *Distribution of mean annual wind speed (left) and wind power density (right) in offshore areas of China with horizontal resolution of 1 km × 1 km and at a height of 100 m*
(PROVIDED BY CMA WIND AND SOLAR ENERGY RESOURCES CENTER)

of the National Development and Reform Commission, under current technology standards and without consideration of the long-distance transmission costs and the resource and environmental benefits of wind power, the electricity price of wind power is 0.20 to 0.25 RMB/kWh higher than that of coal power. If the resource and environmental benefits of wind power factored in to the price for the comparison with coal, the electricity price of wind power is equal to that of coal power. If the increased cost of integrating wind power into the grid and accommodating it, as well as transmitting it long-distance are also factored in, the total costs of wind power in various areas are distinctly different, varying from about 0.05 to 0.3 RMB/ kWh. It is anticipated that the electricity price and cost of wind power will be equal to that of coal power around 2020, even without considering the possibility of a fossil energy resource tax (or environmental tax, carbon tax, etc.). The price of wind power will descend below that of coal power after 2020, even with wind power accommodation and long-distance transmission.

The China Wind Energy Development Roadmap for 2050 involves calculation of wind power supply curves of six onshore wind power bases and the Jiangsu offshore wind power base with a capacity over 10 GW for 2020 and 2030 (Figure 13.10 and 13.11). The results were achieved using the following calculation: on-grid price (capital with interest price) is determined annually for power generation, as is internal rate of return, unit installation cost, etc.; the total cost price is defined as the sum of on-grid price, local access cost, and inter-provincial power transmission cost; the capital with interest price and actual installed capacity of grid unit per km^2 is calculated using Geographic Information Systems (GIS) in combination with economic evaluation analysis, based on geomorphic features, ground observation, remote sensing data, and grid distribution; and then geospatial supply curves (GSC) are drawn for the seven large wind power bases. The GSCs reveal that western Inner Mongolia has the largest technical potential, due to flat terrain, good development conditions for wind power, and lower development cost, while the cost of offshore wind power development is unsurprisingly high in Jiangsu, offering poorest technical potential. Using current wind resource evaluations and wind development technology, if the highest on-grid price is restricted to 0.55 RMB/ kWh (or the total cost price is restricted to 0.7 RMB/ kWh), the economic potential of wind power in seven large-scale bases can only reach an installed capacity of 700 GW.

Estimates indicate that the accumulated installed capacity of wind power will reach up to 200 GW by 2020. If inter-provincial power transmission is not included, the cost of wind power can equal that of conventional energy resources (coal power), the installed capacity can account for 11% of

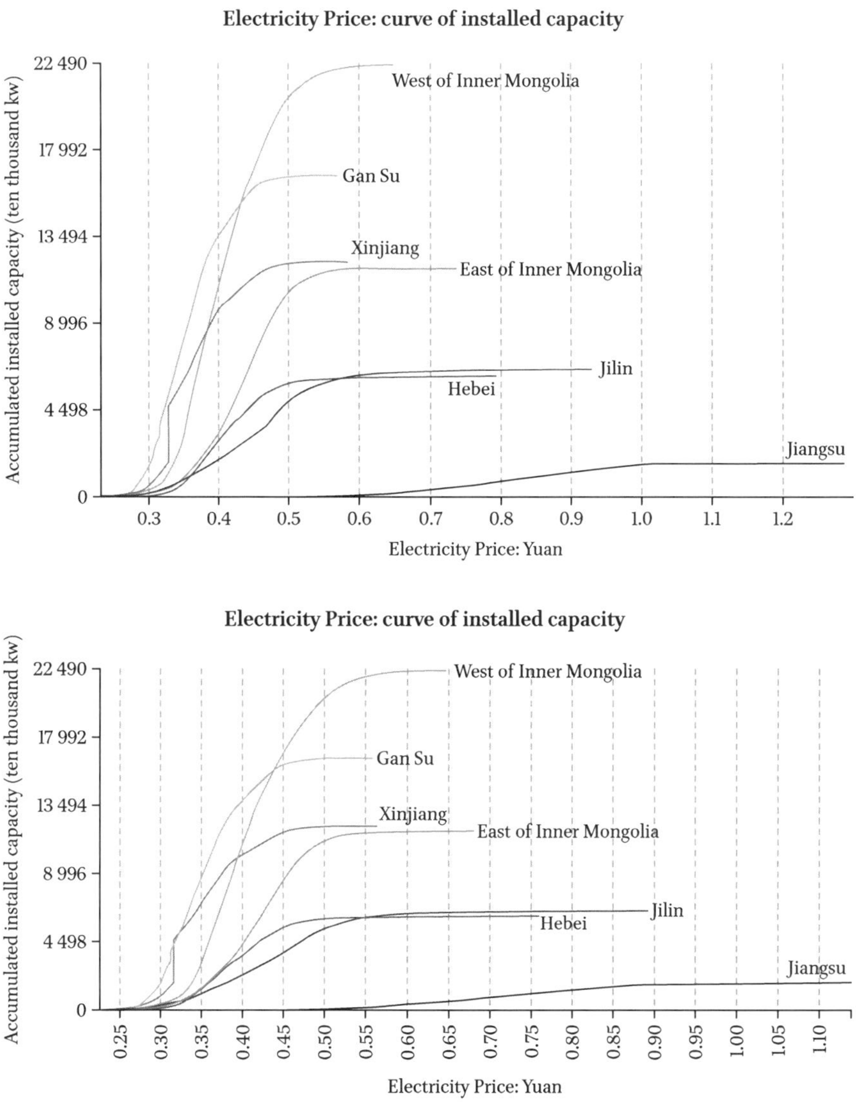

FIGURE 13.10 *Geospatial supply curves (excluding access and transmission cost) of seven large-scale wind power bases in 2020 (top) and 2030 (bottom)*
(DATA SOURCE: CHINA WIND ENERGY DEVELOPMENT ROADMAP 2050 BY THE ENERGY RESEARCH INSTITUTE OF NATIONAL DEVELOPMENT AND REFORM COMMISSION AND IEA)

total energy capacity, and wind power generation can reach 5% of electricity demand. By 2030, the accumulated installed capacity of wind power will exceed 400 GW, accounting for 8.4% of the nationwide power generation, and

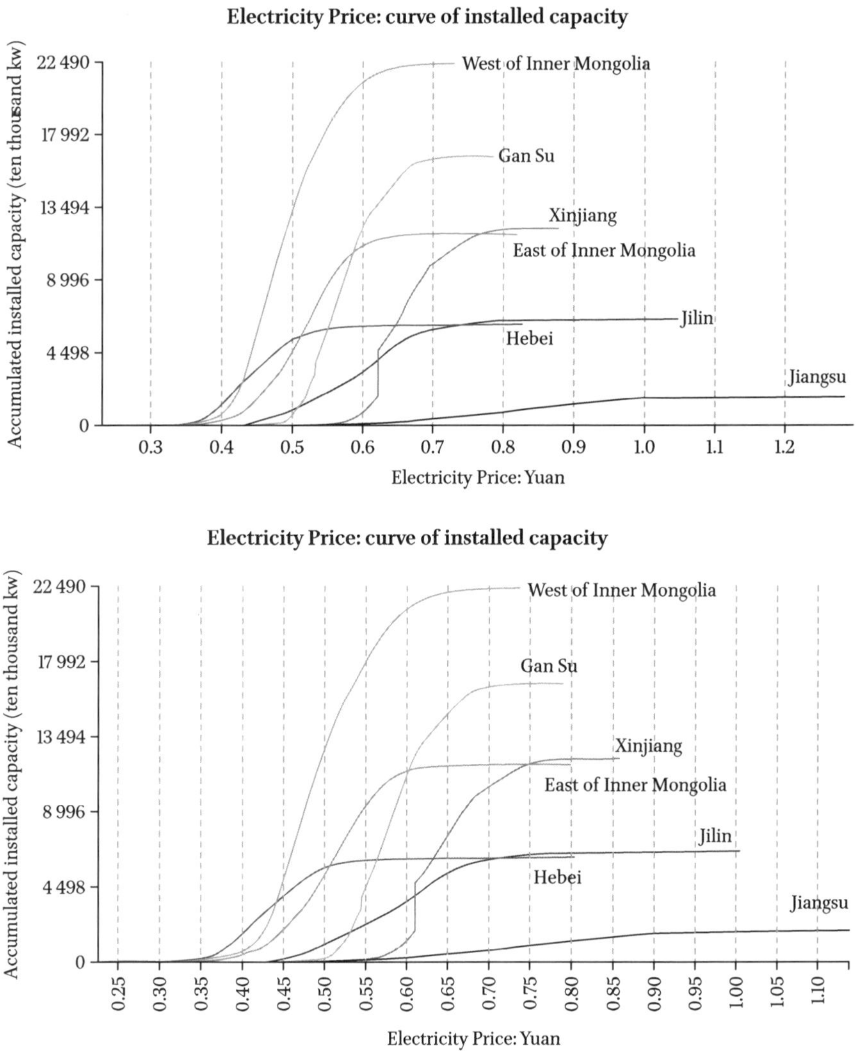

FIGURE 13.11 *Geospatial supply curves (including access and transmission cost) of seven large-scale wind power bases in 2020 (top) and 2030 (bottom)*
(DATA SOURCE> CHINA WIND ENERGY DEVELOPMENT ROADMAP 2050 BY THE ENERGY RESEARCH INSTITUTE OF NATIONAL DEVELOPMENT AND REFORM COMMISSION AND IEA)

about 15% of the power supply. The role of wind power in meeting the power demand, improving the energy structure, supporting the national economy, and furthering social development will be increasingly evident. By 2050, wind power can account for 17% of the power supply of the whole country, with an

TABLE 13.2 *Anticipated investments in wind power in China (constant price in 2010)*

		2010	2020	2030	2050
Electricity price (yuan/kWh) (without consideration of long-distance power transmission and power storage costs)	Onshore	0.57	0.51	0.48	0.45
	Offshore	0.77–0.98	0.77	0.6	0.54
	High sea		>2	2	1
Total investment in that year (RMB 100 million)		1234	1459	2982	4276
Accumulated total investment (RMB 100 million)		3131	17726	38338	120962

Data source: China Wind Energy Development Roadmap 2050 by the Energy Research Institute of National Development and Reform Commission and IEA.

accumulated installed capacity of 1000 GW, making up about 26% of the power supply structure. Wind power will become one of the China's main power supplies and be widely used in industries and in other fields. According to these strategic wind power development objectives, the accumulated investment in wind power in China will reach RMB 1,200 billion by 2050 (Table 13.2).

After 2006, China implemented regional feed-in tariff systems for wind power, and stipulated that the renewable energy development fund will pay for the portion of wind power price that is higher than the benchmark price of desulphurized coal combustion. A price support standard for wind power grid-integration of 0.01 to 0.01 RMB/kWh was decided, dependent upon the distance between wind farm and existing power transmission line. The on-grid electricity price of onshore wind power will equal the benchmark electricity price of desulphurized coal combustion around 2020. The on-grid electricity price support required by wind power will be gradually increased over the next 10 years, reaching its peak value around 2015 and gradually decreasing until 2020. The accumulated on-grid price support for wind power from 2011 to 2020 will reach over 210 billion yuan (Figure 13.12).

If the installed capacity of wind power reaches up to 200 GW in 2020, 400 GW in 2030, and 1000 GW in 2050, and the estimated annual coal power emissions coefficient matches at 751 g/kWh, 727 g/kWh, and 704 g/kWh respectively, the corresponding annual carbon dioxide emissions reduction will be 300 million tons, 600 million tons and 1.5 billion tons, and the corresponding

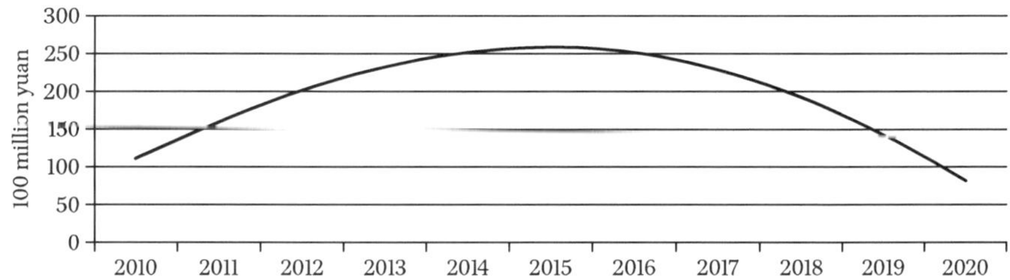

FIGURE 13.12 *Anticipated cost of wind power price support*
(DATA SOURCE: CHINA WIND ENERGY DEVELOPMENT ROADMAP 2050 BY THE ENERGY RESEARCH INSTITUTE OF NATIONAL DEVELOPMENT AND REFORM COMMISSION AND IEA)

annual sulfur dioxide emissions reduction will be 1.1 million tons, 2.2 million tons, and 5.6 million tons.

B *Solar Energy Distribution and Development Potential*

China began investigation of solar energy resources relatively late, so that analysis of energy development potential results looks only at theoretical potential. The nationwide solar energy resource assessment completed by CMA in 2010 suggests that China's solar energy theoretical potential follows a trend that increases, decreases, and increases yet again going from the northwest to the southeast. The theoretical potential of solar energy in the west is generally higher than in the east, as a result of high plateaus vs. plains, inland areas vs. coasts, and arid vs. humid regions (Figure 13.13). The annual solar radiation averaged across the whole country is about 1,500 kWh/m^{2,} broken down to over 1,000 kWh/m^2 in most areas (over 98%) and over 2,000kWh/m^2 in the other 2%. In the south of Tibet and the Qinghai Golmud area receives the greatest amount of horizontal solar radiation in China, topping at 2,140 kWh/m^2 in the south of Tibet, while the low value is found in Chongqing at 905kWh/m^2. Table 13.3 provides the main areas in China distributed by solar energy resources divided into four levels. The "abundant zones" cover the largest area, the "most abundant zone" is basically equal to the area of the "good zone," and the "ordinary zone" only accounts for 3.3% of land area. These results conclude that most areas in China are suitable for solar energy development taking into account only theoretical potential.

Direct solar radiation on a horizontal plane reflects available solar energy for photovoltaic power generation. Figure 13.14 shows the spatial distribution of mean annual horizontal direct solar radiation in China from 1961 to 2008, with a nationwide average value of about 670 kWh/m^2. The figure

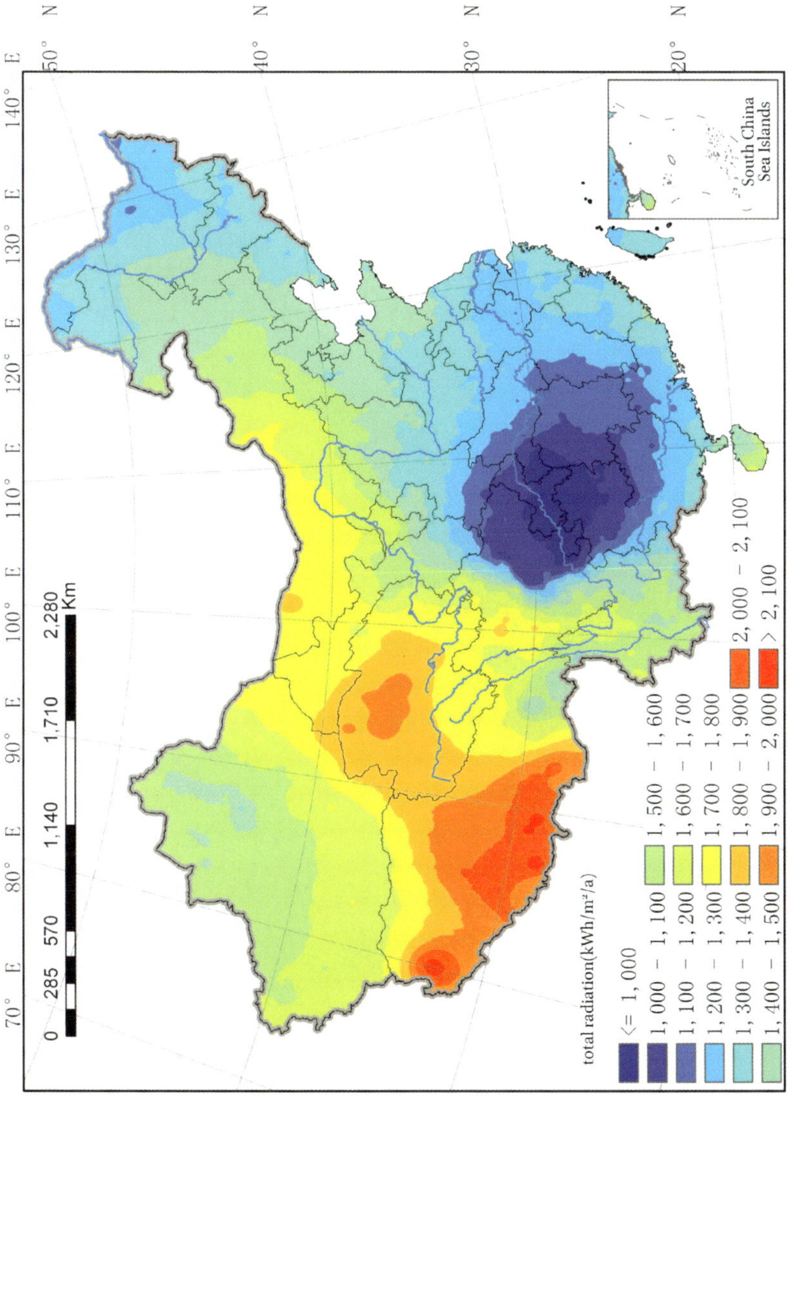

FIGURE 13.13 *Spatial distribution of mean annual solar radiation in China from 1961 to 2008*
(DATA SOURCE: CMA WIND AND SOLAR ENERGY RESOURCES CENTER)

TABLE 13.3 *Division by theoretical solar potential in China*

Levels	Threshold value of total annual amount (MJ/m^2)	Threshold value of total annual amount (kWh/m^2)	Threshold value of average annual radiation (W/m^2)	Occupied area (%)	Main areas
The most abundant zone	≥6300	≥1750	About ≥200	About 22.8	Western Ejinaqi of Inner Mongolia, western Gansu and Jiuquan, most areas we of 100° E in Qinghai, most areas west of 94° E in Tibet, fringe areas in eastern Xinjiang, and parts of Sichuan and Ganzi
Abundant zone	5040~6300	1400~1750	About 160~200	About 44.0	Most areas of Xinjiang, mos parts in western Ejinaqi of Inner Mongolia, western Heilongjiang, western Jilin, western Liaoning, most are of Hebei, Beijing, Tianjin, eastern Shandong, most areas of Shanxi, Ningxia, most areas east of Gansu ar Jiuquan, fringe areas in eastern Qinghai, east of 94° in Tibet, central and wester Sichuan, most areas of Yunnan, and Hainan
Good zone	3780~5040	1050~1400	About 120~160	About 29.8	North of 50° N in Inner Mongolia, most areas of Heilongjiang, eastern centr in Jilin, central and eastern Liaoning, central and western Shandong, souther Shanxi, central and southe Shanxi, fringe areas east of Gansu, central Sichuan,

Levels	Threshold value of total annual amount (MJ/m²)	Threshold value of total annual amount (kWh/m²)	Threshold value of average annual radiation (W/m²)	Occupied area (%)	Main areas
					fringe areas in eastern Yunnan, southern Guizhou, most areas of Hunan, most areas of Hubei, Guangxi, Guangdong, Fujian, Jiangxi, Zhejiang, Anhui, Jiangsu, and Henan
Ordinary zone	<3780	<1050	About <120	About 3.3	Eastern Sichuan, most areas of Chongqing, north central Guizhou, west of 110° E in Hubei, and northwestern Hunan

Data source: CMA Wind and Solar Energy Resources Center.

demonstrates how the spatial distribution of annual mean direct horizontal solar radiation is similar to the total radiation in China, with distribution west to east as follows: most areas of Xinjiang are below 1,000 kWh/m^2, and the hinterland of Xinjiang's Taklimakan Desert has even lower radiation at about 600 kWh/m^2; central and western Inner Mongolia, western Gansu, and most areas of Qinghai and Tibet have a high horizontal direct solar radiation, with an annual total over 1,000 kWh/m^2, and broken down western and southern Tibetan areas reach 1,300 kWh/m^2 and some even reach the maximum level of 1,530 kWh/m^2; eastern Sichuan, Chongqing, Guizhou, central and northern Guangxi, Hunan, and southwestern Hubei have low horizontal direct solar radiation with an annual levels below 500 kWh/m^2, from which parts of Chongqing reach only less than 300 kWh/m^2 including the minimum level of 240kWh/m^2; central and southern China and southern areas range from 500 to 700 kWh/m^2, including areas such as Henan, southern Shandong, Anhui, Jiangsu, Shanghai, Zhejiang, most areas of Hubei, Jiangxi, Fujian, Guangdong,

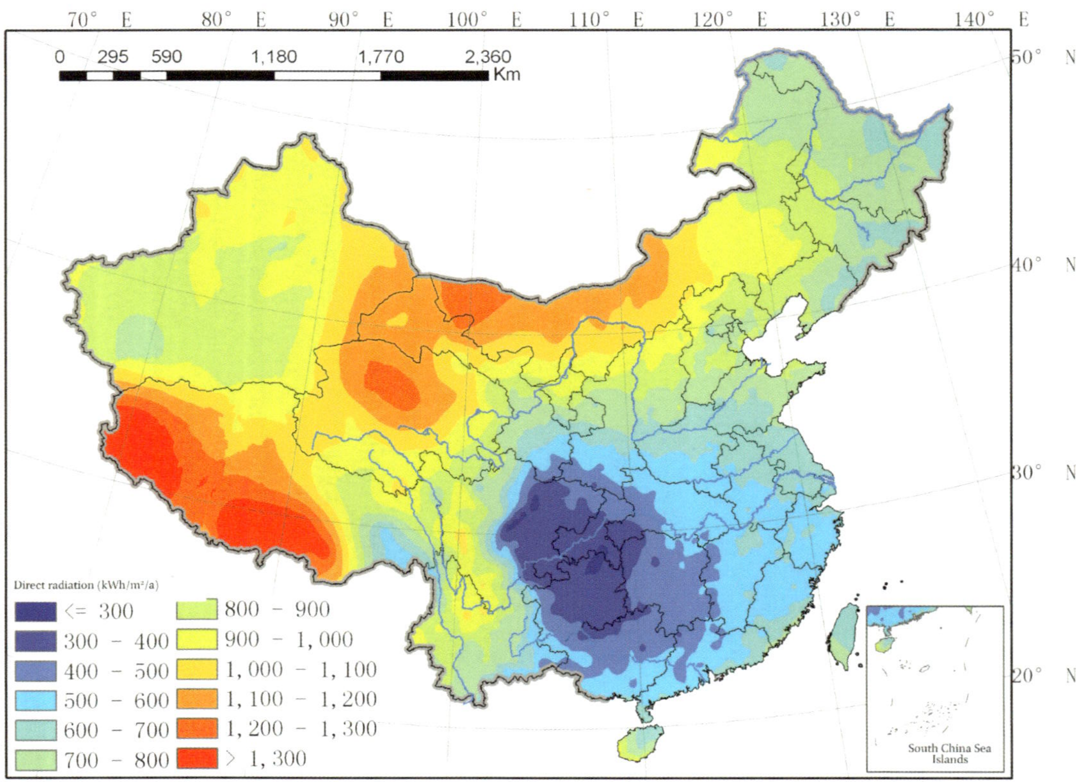

FIGURE 13.14 *Spatial distribution of mean annual horizontal solar direct radiation in China from 1961 to 2008* (DATA SOURCE: CMA WIND AND SOLAR ENERGY RESOURCES CENTER)

southern Guangxi; and North China ranges from 700 to 1000 kWh/m², in areas such as Shanxi, Hebei, Beijing, Tianjin, eastern Inner Mongolia; Liaoning, Jilin, Heilongjiang, and most areas of northeastern Inner Mongolia have levels at about 800 kWh/m².

Synthesizing and summing the data, the theoretical potential of horizontal solar energy resource reaches 1,680 billion kW, and that of horizontal direct solar radiation resources is 920 billion kW. Xinjiang has the greatest horizontal solar theoretical potential energy of all provinces and autonomous regions, at 285.523 TW, however Tibet has the greatest annual total theoretical potential energy at 1885.50 kWh/m².

Global photovoltaic power generation has exceeded 40 GW, and the growth of photovoltaic power generation in the EU has exceeded the growth of wind power, which makes photovoltaic power generation the third largest renewable energy in term of development, second only to hydroelectric and wind power. The 12th Five Year Plan for Renewable Energy Development (Exposure Draft) of the National Energy Administration (NEA) demands the implementation of detailed solar energy resource assessments, creation of a diversified solar energy power generation market through policy support and guidance, and comprehensive consideration of solar energy resources, construction conditions, development costs, and electricity market demands. By 2015, the accumulated installation of solar water heating systems will total 400 million m² and the installed capacity of solar energy generation will reach 10 GW. The installed capacity of large-scale grid-integrated power plants is 5 GW, that of urban-rural grid-integrated power plants is 3 GW, and that of off-grid and distributed generation is 2 GW, with key regions including Qinghai, Gansu, Inner Mongolia, Tibet, Ningxia, Yunnan, and Hainan. The plan proposes that the total installed capacity of solar energy shall reach 1 GW. The Research Report on China's 12th Five Year Plan for the Electricity Industry of the China Electricity Council proposes to 1) give full advantage to solar energy photovoltaic generation for scattered power supply, and promote and encourage consumers in remote areas to adopt photovoltaic generation systems or construct small-scale photovoltaic power plants to resolve power supply problems or to bring electricity to people without access to power, especially in key provinces, autonomous regions, and cities like Tibet, Qinghai, Inner Mongolia, Xinjiang, Ningxia, Gansu, and Yunnan. 2) Equip urban buildings and public facilities with auxiliary solar energy photovoltaic power generation devices, in order to increase the application of urban renewable energy and provide a wider market for solar energy photovoltaic power generation. 3) Develop rooftop photovoltaic power generation in urban cities such as Beijing and Shanghai, and in cities in Jiangsu, Guangdong, and Shandong. 4) Establish large-scale grid-integrated solar energy photovoltaic power plant demonstration projects at Dunhuang in Gansu, Chaidamu Basin in Qinghai, and Lhasa in Tibet, and establish solar thermal power plant demonstration projects on spare lands, such as the desert, the Gobi, barren beaches, etc. in Inner Mongolia, Gansu, Qinghai, and Xinjiang.

According to the forecast from the Research Report on the Optoelectronics Industry, future investment will flow to the integration of photovoltaic solar energy with buildings, concentrated photovoltaic (CPV), and concentrated solar thermal power generation. China's 12th Five Year Plan for Photovoltaic Solar Energy Industry Development suggests that by 2015 the cost of photovoltaic systems will drop to 15,000 RMB/kW and that the cost of power generation will decrease to 0.8 RMB/kWh, and further that fair on-grid price will be achieved at power distribution. By 2020, the cost of the system and of power generation will drop to 10,000 RMB/kW and 0.6 RMB/kWh respectively, and fair on-grid price will be achieved at power generation, making the industry competitive in primary electricity markets. The concentrated solar thermal power (CSP) development roadmap launched by the IEA in 2011 predicts that CSP will be able to meet 11.3% of global electricity demands with a potential industrial scale of 320 billion RMB by 2050. According to statistics from Huatai United Securities, the installed capacity of thermal solar energy power generation in China will reach nearly 3,000 MW before 2015. The total market share is calculated to reach 45 billion RMB factoring comparable cost of photovoltaic systems.

III Safeguard Policies for Wind and Solar Energy Development

On December 26, 2009, Amendments to the Renewable Energy Law of the People's Republic of China were passed by the 12th session of the Standing Committee of the 11th National People's Congress, and they were put into effect April 1, 2010. The amendments alter Article 6 in the Renewable Energy Law of the People's Republic of China passed in 2005, legally restraining the development of excessive new energy industries in the short-term. The amendments to the Renewable Energy Law of the People's Republic of China emphasize the role of the central government in overall planning. They explain that state electricity regulation organizations and financial departments of the State Council shall establish annual renewable energy resource purchasing indicators in accordance with overall plans, and declare that grid companies must reach the minimum indicator. They are the first legal documents to mention smart grid development plans, to propose the establishment of a renewable energy development fund by from state financial authorities, and to specify that the fund will be used to support "renewable energy resource utilization projects in rural and pastoral areas, construction of independent renewable energy power systems in remote areas and sea islands, exploration of renewable energy resources, assessment and construction of information system, promotion of localized renewable energy development and equipment production." The amendments to the Renewable Energy Law of the People's Republic of China facilitate a healthy and orderly development for the new energy industry and for smart grid construction, improve grid companies' market incentives for purchasing renewable energy, and further strengthen the development of renewable energy in China.

In order to guarantee smooth grid-integration for wind power and the safe operation of the power system, the NEA launched the *Interim Measures for Wind Farm Power Forecasting Management* in 2011, establishing the following requirements. All grid-integrated wind farms shall be capable of forecasting wind power and shall carry out forecasts as per specific requirements. The grid scheduling organization shall arrange power generation plans of wind farms in accordance with power forecasting results reported by wind farms to ensure secure grid operation. Newly constructed wind farms shall simultaneously establish wind power forecasting systems and develop mechanisms for power generation planning and reporting. All wind farms' forecasting systems were officially put into operation on July 1, 2012. To aid in the implementation of the "Interim Measures for Wind Farm Power Forecasting Management," the NEA published the *Implementation Rules of Wind Power Forecasting and Coordinated Grid Operation* (*Trial*) in 2012, establishing the following requirements. The CMA shall be responsible for establishing numerical values for the wind power forecasting service platforms and creating a system to guarantee business operations to provide numerical weather prediction and supportive wind power forecasting technology to the public. Wind power development companies shall be responsible for the wind farm's power generation forecast. The grid scheduling organization shall be responsible for the power system's wind power forecast, and shall build power scheduling operations mechanisms based on wind power forecasts to guarantee the prior scheduling of wind power and the implementation of protective wind power purchasing measures. The implementation of these measures taken by the NEA for the promotion of wind power shall effectively relieve the instability of on-grid wind power scheduling, and shall promote the construction of smart grids.

Wind energy resources are widely distributed throughout China. The three-norths and coastal areas are suitable for large-scale wind energy resource development. Due to abundant wind energy resources and scattered distribution, mountain regions on islands are suitable for small-scale scattered development. The NEA published the notice on scattered access to wind power development in 2011 based on international experience to achieve centralized development of large-scale wind farms and to positively and appropriately explore development of scattered wind power access. Wind turbine units shall be prepared and have access to the power distribution system in operation, new transformer substations shall not be constructed for wind power access, and wind power transmission by boosting inverters shall not considered. Grid companies shall measure and purchase all wind power generated by the scattered multipoint access system. This measure taken by NEA supports the improvement of low-speed wind farm development in China's inland areas and effectively relieves difficulties in wind power grid-integration.

Since 2009, China has enhanced its solar energy development and utilization policy and implemented photoelectric solar energy buildings demonstration projects, golden sun demonstration projects, and concessions bidding, which facilitate lower cost per kW for photovoltaic systems and lower cost of power per kWh. The application of photovoltaic technology on buildings and the power plant pilot in the desert greatly increase the ratio of grid-integrated solar power systems. Germany has had the greatest success with photovoltaic development by means of fully formed feed-in tariffs and cost sharing measures. All photovoltaic power generation projects adopt regionally variant feed-in tariffs, and the extra cost incurred from photovoltaic power generation is covered by the grid, and is integrated into the support scope of the national renewable energy fund. This method leads the consumers of public power utilities support photovoltaic power, lends greater attention to photovoltaic power generation, and stimulates rapid development of the industry. On August 1, 2011, the National Development and Reform Commission released the notification for improving the on-grid electricity price policy of photovoltaic solar energy power generation, which is crucial to opening the market to photovoltaic application in China. Photovoltaic solar energy power generation projects which got approval for construction before July 1, 2011, are completed and put into operation by December 31, 2011, and for which the price of power was not approved by the National Development and Reform Commission, will have an on-grid electricity price established as a unified price of 1.15 RMB/kWh as per this new notice. Photovoltaic solar energy power generation projects that got approval on and after July 1, 2011, and that got approval before July 1, 2011 but are not completed and operating by December 31, 2011, will have an on-grid electricity price of 1 RMB/kWh in all provinces, autonomous regions, cities, excluding Tibet, where the on-grid electricity price remains 1.15 RMB/kWh. The price will be appropriately adjusted in the future in accordance with changes to investment cost, technological advancements, etc. The 12th Five Year Plan for National Economic and Social Development of the People's Republic of China proposes opening diversified solar energy generation markets, promoting the construction of large-scale power plants in deserts, constructing scattered photovoltaic roofs on buildings, establishing micro-grid systems for photovoltaic projects, and promoting the construction of photovoltaic generation in areas without access to power.

(This article was originally published in Chinese in 2012.)

CHAPTER 14

Meteorological Disaster Trends and Impacts in China

Gao Ge, Zhao Shanshan and Xu Ying

Abstract

Major disastrous weather and climate in China gone through some unquestionable changes in location and duration over the past fifty year. Changes to drought events, for example, vary by region. Increased frequency of droughts is found in the area stretching from the Liao River Plain, Hai River Plain, Loess Plateau, and Sichuan Basin to the Yunnan-Guizhou Plateau. The number of annual precipitation days has risen as a whole throughout the country, with marked increase in the south and a strong decrease in the north. The frequency of tropical cyclones seems to have decreased although the average intensity of yearly landings has increased. Very high summer temperatures were more frequent, leading to an increase in annual labeled "hot days" throughout the country. The number of days with disasters resulting from low temperatures, frost, and sand and dust storms has experienced a decrease. For nearly 20 years, the direct economic losses caused by the meteorological disasters have increased, although death toll has decreased greatly. Apart from the impact of disastrous weather, extreme climate, and weather changes, social and economic development has also contributed to greater vulnerability of human society and to greater risk of meteorological disaster. The major meteorological disasters that occurred over the last decade have left a profound impact on human society due to emerging climate change and social and economic development. It is estimated that the meteorological disasters and secondary disasters will continue to increase over the next 10 to 50 years in China. Therefore, adaptation to climate change should also include improvements to disaster prevention and scientific measures that aim to reduce or avoid risks.

Keywords

disastrous weather and climate – meteorological disaster – influence – trends

* Gao Ge, National Climate Center, China Meteorological Administration, senior research engineer, Ph.D., studies climate change impacts on water resources, meteorological disaster

 | DOI 10.1163/9789004274648_015

A meteorological disaster is a disaster caused directly or indirectly by meteorological factors, and which brings about losses to human life or economic damage. With eastern China located in the East Asian Monsoon Region and western China landlocked, the country is left strongly influenced by the most serious meteorological disasters due to its diverse topography, complex weather, and climate system. The average direct economic losses caused by China's meteorological disasters annually, which include a variety of disasters and occur at a high frequency, for long periods of time, across scattered areas, at various times throughout the year, and which have a wide range of impact and cause serious damage, account for over 70% of the losses from all the natural disasters. With the changing climate and rapid growth of the economy, new patterns and characteristics have also appeared in the behavior of meteorological disasters over the last few decades, which severely affect the sustainable development of human society and the economy, creating new challenges for disaster prevention and reduction as well as for adaptation to climate change.

I Temporal and Geographical Changes to Major Weather and Climate Disasters in China over the Last 50 Years

For nearly 50 years, the major weather and climate disasters in China have undergone evident changes seasonally and geographically, which has had a direct impact on the area affected by the disaster, dealing with altered frequency, scope, duration, and intensity of meteorological disasters.

A *Drought*

Changes to drought patterns have been roughly identical to changes in precipitation pattern the last 50 years. East of 110°E, the precipitation increased in the south and decreased in the north, while west of 110°E, precipitation tended to increase in the middle-west but decreased in the eastern portion. These patterns inflict blunt changes on regional droughts and floods. From 1958 to 2007, a zone of relatively severe drought stretching from the Liao River Plain, Hai River Plain, Loess Plateau, and Sichuan Basin to the Yunnan-Guizhou Plateau was formed, where droughts happened more often and were more severe. This

monitoring, risk assessment, etc. Zhao Shanshan, senior engineer, Ph.D. conducts research on meteorology disaster monitoring and impact assessment. Xu Ying, Ph.D., researches causes of climate change, detection, and future climate change assessment, climate disaster risk assessment.

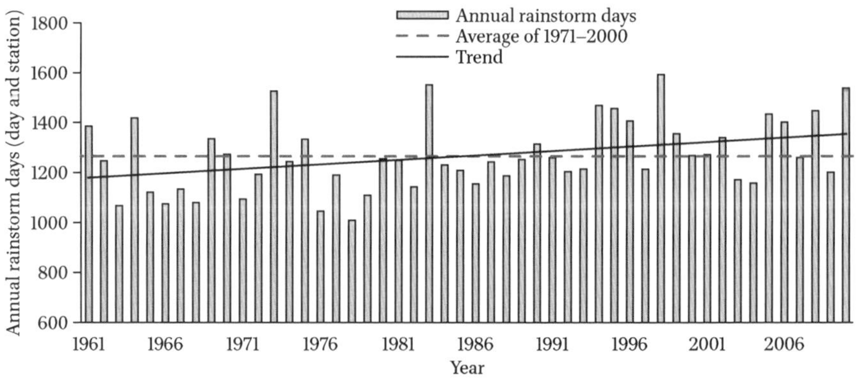

FIGURE 14.1 *Observed annual number of rainstorm days in China from 1961 to 2010*

trend is interlinked with the decrease of precipitation, and with global warming, which serves as another main exacerbating factor to the trend of increasing drought in these areas.[1]

B *Rainstorm*

From 1961 to 2010, the annual number of rainstorm days reported by meteorological stations throughout the country increased and rainstorm events were also on the rise. The number of rainstorm days in 2010 was the third highest since 1961, raking only behind 1998 and 1983 (as shown in Figure 14.1). Both precipitation and rainstorm days tended to increases in southeastern China, which is one of the main contributors to frequent floods. The number of rainstorm days decreased in most northern river basins in China,[2] but nevertheless increased in some areas.

C *Tropical Cyclones*

From 1961 to 2010, the number of tropical cyclones seems to have decreased in the northwest Pacific Ocean and the South China Sea, and the ones

1 Lu Guihua, Yan Guixia, Wu Zhiyong, and He Hai 陆桂华、闫桂霞、吴志勇、何海, "Jin wu shi nian lai zhong guo gan han hua te zheng fen xi 近50年来中国干旱化特征分析 [Analysis on draught characteristics within the last 50 years in China]," *Shuili shuidian jishu* 41 (2010): 78–82.

2 Chen Yu, Chen Xianyan, and Ren Guoyu 陈峪、陈鲜艳、任国玉, "Zhong guo zhu yao he liu liu yu ji duan jiang shui bian hua te zheng 中国主要河流流域极端降水变化特征 [Variation of extreme precipitation over major river basins in China]," *Qihou bianhua yanjiu jinzhan* 6 (2010): 265–269.

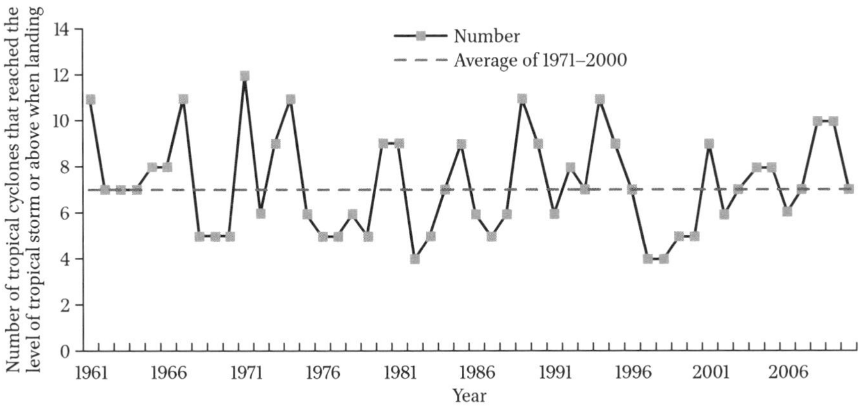

FIGURE 14.2 *Number of tropical cyclones that reached the tropical storm level or above when landing (1961 to 2010)*

number that reach the tropical storm level or higher when landing also tended to decrease. The number was smallest in 1997 and 1998, and has increased markedly since then (as shown in Figure 14.2). The average annual landing intensity of the tropical cyclones has been getting consistently stronger for a long time if judged by observing the central lowest sea-level pressure and proximal maximum wind speeds. The intensity of the tropical cyclone landings in both the southern part of China and eastern China have increased, with landings more concentrated on specific coastal zones and at specific time period.[3] During the 35 year period from 1975 to 2009, the mean annual onshore duration for tropical cyclones has increased, which has been correlated to the increase in tropical cyclone precipitation and the large-scale changes to the steering current.[4]

D *High Summer Temperatures*

From 1961 to 2010, hot days with a daily maximum temperature greater than or equal to 35° during the summer (from June to August) greatly increased throughout the country, especially during the summer of 2010 when the

3 Yang Yuhua, Ying Ming, and Chen Baode 杨玉华、应明、陈葆德, "Jin 58 nian lai deng lu zhong guo re dai qi xuan qi hou bian hua te zheng 近58年来登陆中国热带气旋气候变化特征 [The climate change characteristics of tropical cyclones landing in China over the past 58 years]," *Qixiang xubao* 67 (2009): 689–696.

4 Chen Xiaoyu, Wu Liguang, and Zhang Jiaoyan, "[Increasing duration of tropical cyclones over China]," *Geophysical Research Letters* 38 (2011), L02708, doi:10.1029/ 2010GL046137.

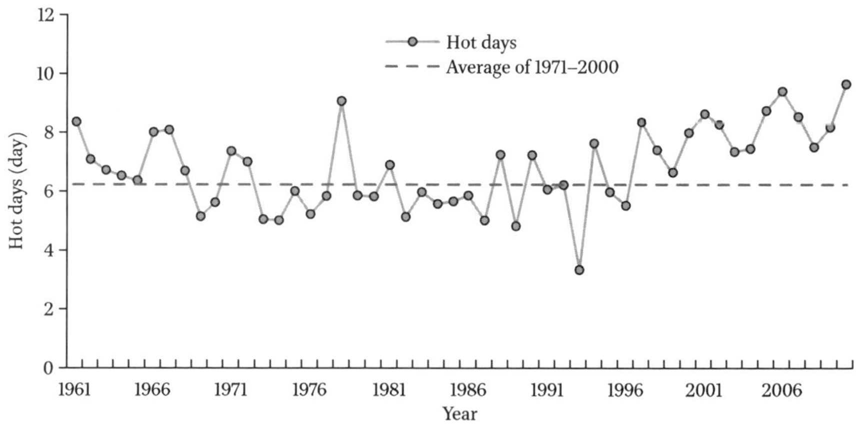

FIGURE 14.3 *Number of hot days with the daily maximum temperature greater than or equal to 35° during summer in China from 1961 to 2010*

average number of hot days throughout the country was 9.7, equal to the historical maximum for the same period which took place in 1961 (as shown in Figure 14.3). Since 1997, the number of hot days in summer have remained above average.

E *Cold Damage, Frost, and Snowfall*

Cold damage is one of the most sever agricultural meteorological disasters in China, including spring and autumn cold damage in the south of China and summer cold damage in northeastern China, which have a significant impact on the grain output. Due to global warming, the number of days causing cold damage or with frost and snow have decreased.

The frequency of spring cold damage in the south of China has occurred infrequently since 2000, but most areas suffered relatively more severe cold damage in the 1990s.[5] There was relatively more cold damage during the early rice-seeding period in the southernmost part of southern China from 1951 to 1970, but less from 1971 to 1984. The greatest cold damage happened in the period from 1985 to 2000, but it seems to have decreased after 2000. The frequency of cold damage events in the northern part of southern China was high

5 Han Rongqing, Chen Lijuan, Li Weijing, and Zhang Peiqun 韩荣青、陈丽娟、李维京、张培群, "2–5 yue wo guo di wen lian yin yu he nan fang leng hai shi kong te zheng 2–5月我国低温连阴雨和南方冷害时空特征 [The spatial and temporal distribution characteristics of China continuous cold rainy weather and cold damage in the South from February to May]," *Yingyong qixiang xuebao* 20 (2009): 312–320.

during 1969 and 1992, but low during other periods, with little cold damage from 1997 until the present. There were relatively more cold damage events in the middle and lower reaches of the Yangtze River from the middle of the 1960s to the early 1970s and from 1980 to 1999, while relatively fewer events occurred from the end of the 1950s to the early 1960s, during the mid-1970s and after 2000. The Sichuan Basin suffered more from frequent cold damage from 1983 to 1999, but less from 1964 to 1970, 1980 to 1984, as well as after 2000.

Since the 1980s, the frequency of cold wind forming dew on rice in southern China tended to decrease, especially after the late 1990s.[6]

Due to global warming, the summer cold damage in northeastern China also been decreasing.

From 1961 to 2010, the average number of snow days and frost days annually has starkly decreased in China. The average date of latest frost has gotten progressively earlier across the country since the 1980s onward, while the date of the first frost has gotten progressively later since the 1990s, and the average frost-free days nationwide has grown since the 1980s.[7]

F *Sand and Dust Storms*

Sand and dust storms throughout northern China are annually declining on average. From the middle 1980s onwards, the number of days with sand and dust storms was generally lower than average, declining rapidly. Since 2003, all the average number of sand and dust storm days annually in the north was less than one. Despite the decreasing trend of sand and dust weather in the north over nearly half a century, some areas, like northern Qinghai, western Xinjiang, and Xi Lin Hao Te in Inner Mongolia, have experienced an increase in sand and dust weather. The strong sandstorms typical in northern China have been decreasing in a fluctuating pattern for nearly half a century, with a high concentration of sandstorms in the 1950s, and fewer in the 1990s, but a relative increase has been seen during the early 21st century.[8]

---

6 China Meteorological Administration 中国气象局, Zhong guo zai hai xing tian qi qi hou tu ji (1961–2006) 中国灾害性天气气候图集（1961–2006年）[Atlas of China Disastrous weather and Climate (1961–2006)], (Beijing: Zhongguo qixiang chubanshe, 2007), 67.

7 Ye Dianxiu and Zhang Yong 叶殿秀、张勇, "1961–2007 nian wo guo shuang dong bian hua te zheng 1961–2007 年我国霜冻变化特征 [Characteristics of Frost Changes from 1961 to 2007 over China]," *Yingyong qixiang xubao* 19(2008): 661–665.

8 Zhou Zijiang and Zhang Guocai, "[Typical severe dust storms in northern China during 1954–2002]," *Chinese Science Bulletin* 48 (2003): 2366–2370.

II Influence of Meteorological Disasters

Meteorological disasters accounts for 71% of all natural disasters in China, including rainstorm floods, the severe storm disasters such as thunder, lighting, and hail, and tropical cyclones, which are responsible for the highest casualties, while rainstorm floods, droughts, and tropical cyclones are responsible for greater direct economic losses. From 1990 to 2010, there was an annual average of 210.5 billion RMB in direct economic losses caused by meteorological disasters. Over the last five years, the direct economic losses were consistently above average, and the record economic loss of more than 500 billion RMB happened in 2010 (as shown in Figure 14.4). As the economy develops, the ratio of direct economic losses caused by meteorological disasters to GDP has noticeably decreased. The death toll caused by meteorological disasters averages 4044 people annually, which reveals a general downward trend. Compared to the 1990s, the death toll over the last 10 years has greatly declined, falling below the average (as shown in Figure 14.5). Meteorological disasters have a significant impact on Chinese agriculture, where droughts caused damage to crops over the largest area, accounting for 57% of the total crop damage area, followed by floods with 24%, and then cold damage and snow disasters, accounting for 11%.

Meteorological disasters, especially major meteorological disasters, have a distinct impact on the economy and on human life. As the climate warms, more frequent extreme weather and climate events result in the greater magnitude and frequency of meteorological disasters. As the economy continues

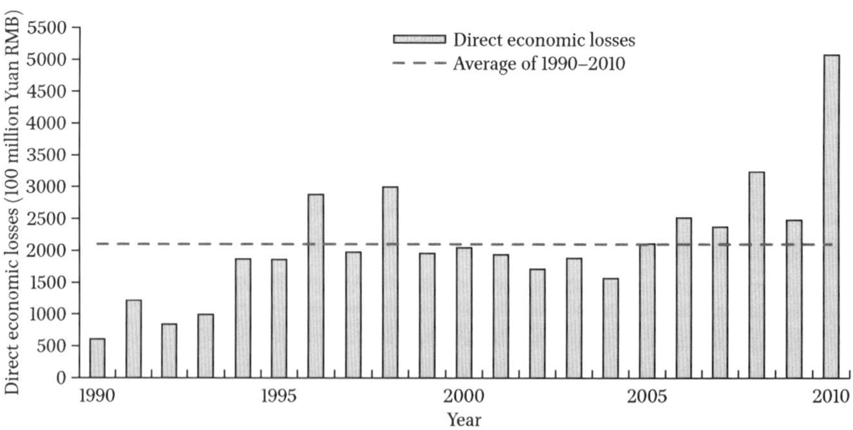

FIGURE 14.4 *Direct economic losses caused by meteorological disasters in China from 1990 to 2010*

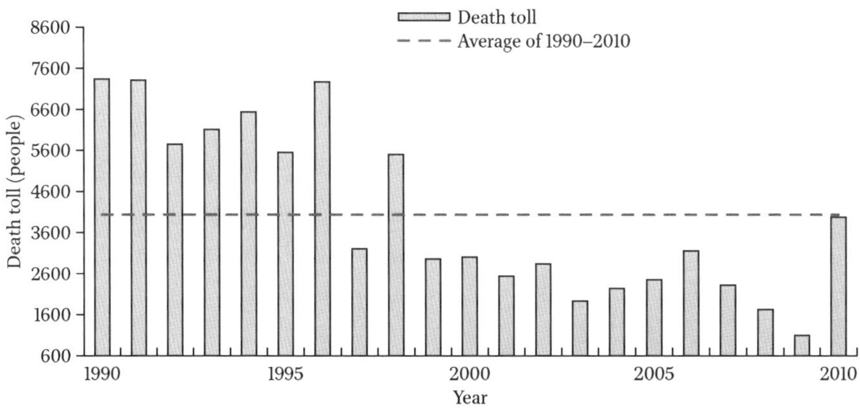

FIGURE 14.5 *Death toll caused by meteorological disasters in China from 1990 to 2010*

to develop and becomes more exposed and vulnerable, the influence of meteorological disasters takes on new forms. Many problems deserve further reflection, for example, the impact of meteorological disasters becomes increasingly intense and affects a larger scope as relationships between various industries get close. Urban development creates high concentrations of population and property, increasing their level of exposure and vulnerability. Higher standards of living and development result in disruptions to the resource environment, further increasing the risks of meteorological disasters. A number of crucial structures such as energy power grids, basic infrastructure, transportation infrastructure, etc. are increasingly likely to be affected by meteorological disasters. Only an in-depth analysis of factors triggering disasters and features that influence meteorological disasters in disaster stricken areas under current conditions of major meteorological disasters could improve China's disaster prevention and reduction measures, allowing it to effectively avoid meteorological disaster risk and better adapt to climate change.

A *Influence of Drought*

An increase in droughts, a drop in water levels of rivers, lakes, and reservoir, and insufficient reservoir storage will negatively impact agricultural production, drinking water availability for humans and animals, city water supplies, and the natural environment. Dry climates will also result in more heat issues and forest fires. Droughts lead to water and electricity shortages, rising prices, and factory closures, imposing a widespread effects on economic development.

The area of crops impacted and affected by droughts throughout the country has been increasing for several decades. Since the beginning of the 21st century, regional droughts have happened more frequently wreaking major havoc. The droughts are extending into the usually rainy regions of the south and east. Northern China experienced extensive severe droughts from spring to summer in 2000 and 2001 creating the greatest two years of drought induced crop loss throughout the country since nearly 60 years. In 2003, some areas in the south of the Yangtze River, southern China and southwestern China underwent severe summer and autumn droughts. In 2004, the western portion of northeast China and eastern Inner Mongolia experienced severe droughts from spring to summer while southern China and the middle and lower reaches of the Yangtze River suffered extensive severe autumn droughts. In 2005, severe droughts stretching from autumn all the way to spring attacked the southernmost portion of southern China. In the summer of 2006, severe summer droughts hit Sichuan and Chongqing provinces. Drought in Chongqing in 2006 was at its worse since at least a century. In 2008, northeastern China and northern China suffered from severe droughts from winter to spring. In 2009, the winter wheat region experienced serious droughts from autumn to winter. From autumn 2009 through March 2010, a very severe drought of intensity rarely seen in history hit southwestern China. In 2011, northern China and the Huanghuai region experienced the worst meteorological drought in the last 41 years, lasting from autumn to winter. During January and May in 2011, a drought unprecedented in 60 years ravaged the middle and lower reaches of the Yangtze River basin from winter to spring, and a summer drought along with high temperatures hit Guizhou, Hunan, and Yunnan provinces.

1 The Unprecedented Drought in Southwest China from Autumn 2009 to Spring 2010

From September 2009 until the middle March of 2010, precipitation in Yunnan, Guizhou, southern Sichuan, and northwestern Guangxi was 30% below average at same period, while precipitation in some areas of central and eastern Yunnan, southwestern Guizhou, and northwestern Guangxi was 50% to 80% below average. Temperatures in most parts of southwest China and Guangxi were 1 to 2° higher than normal for that period. The high temperatures and below average precipitation levels resulted in a continuous drought, which became southwest China's worst drought since records began, stretching from autumn to spring. Drought has severe impacts on agriculture, hydroelectric power generation, shipping, tourism, and human life, affecting over 80 million people in Yunnan, Guizhou, Sichuan, Chongqing, Gansu, Qinghai, Hebei,

Hubei, Hunan, Guangxi, and Guangdong, and causing direct economic losses of over 30 billion RMB. The disaster was worst in Yunnan, Guizhou, Sichuan, and Guangxi, with the direct economic losses exceeding 28 billion RMB.[9]

2 The Worst Drought in 60 Years for in the Middle and Lower Reaches of the Yangtze River Basin in Winter and Spring of 2011

From January to May, 2011, the consistent lack of rainfall in Hubei, Hunan, Jiangxi, Anhui, and Jiangsu, when precipitation averaged only 260.9 mm, which is 51.1% below average, become the lowest historically documented precipitation level for this period since records began in 1951, and resulted in severe droughts in some areas. This drought was sever, long, concentrated in particular areas, and struck during flood season, resulting in a consistently low water level in Dongting Lake, near complete draining of most of the 107 small reservoirs in Lixian county and in Hunan province, and thousands of reservoirs of different size in Hubei with almost no water, leaving humans and livestock in Hubei and Hunan short of drinking water. The surface area of major lakes in Jiangsu province decreased by varying degrees, and Shijiu Lake in Gaochun county was left almost entirely dry. The drought also brought losses to the aquaculture industry and agricultural production.

3 Severe Drought in Southwest China Again in Summer and Autumn of 2011

Since the summer of 2011 (until September 15th), the precipitation level in the five provinces of Guizhou, Guangxi, Chongqing, Yunnan, and Sichuan in southwest China was 527.5 mm, 19.5% below average (655.2 mm) for that period. This was the second lowest precipitation level for that period over the last 61 years, the period with the greatest number of rain free days, and the third longest rain free continuous streak, while the average temperature was the second highest. The precipitation in Guizhou was 362.4 mm, 37.9% below the average for that period, making it the lowest over the last 61 years, and both the number of rain free days and the average temperature ranked highest since 1951, while the number of high temperature days ranked second. Consistent high temperatures and sparse rainfall triggered drought disasters in Guizhou, Yunnan, Chongqing, Guangxi, and Sichuan, averaging 37.6 days of drought in southwest China Guizhou suffered the worst droughts and the greatest number of drought days for that period since 1951 with 42.8 days,

9 China Meteorological Administration 中国气象局, Quan guo qi hou ying xiang ping jia 全国气候影响评价 [National Climate Impact Assessment], (Beijing: Zhongguo qixiang Chubanshe, 2010), 22–23.

followed by Sichuan. High temperatures combined with drought severely impact industry, agriculture, forestry, water resources, hydroelectric power generation, the natural environment, and human life. Other results include stream cutoffs, reservoirs and ponds left dry, a reduction of available water, causing water shortages in some counties, power-supply limits that force some industrial enterprises to suspend operations or slash production, and increased risk of forest fires, which ultimately resulted in a number of town and forest fire in Chongqing and Guizhou. Statistics at the end of August suggest that the population affected by drought reached 38.763 million in Guizhou, Yunnan, Chongqing, Guangxi, and Sichuan; and 3.344 million hectares of crops were damaged, causing about 21.3 billion RMB in direct economic losses.

B *Influence of Rainstorm Flooding*

Over the last 50 years, the area of crops hit by rainstorm floods nationwide increased remarkably The 1990s saw the greatest average area becoming water-logged, with the largest water-logged crop area occurring in 1991 with 24.596 million hectares and the second largest happening in 1998 with 22.292 million hectare. Although water-logged area decreased slightly at the beginning of the 21st century, the water-logged crops in 2003 and 2010 nevertheless ranked the third and fifth largest areas respectively over the last 50 years. Frequent floods occurred in the Huaihe River Basin, where torrential rain in 2003 left more than 58 million people affected in Anhui, Jiangsu, and Henan provinces, causing over 35 billion RMB of direct economic losses. In 2007, the entire Huaihe River flooded severely, a flood second only to the one in 1954, leaving more than 26 million people affected, over 30 people dead, and leading to more than 17 billion RMB of direct economic losses. In the early summer of 2008, the Pearl River Basin and the upper reaches of the Xiang River suffered from a severe flood; which triggered landslides and debris flows, affected 36 million people, killed 177 people, and led to 29.66 billion RMB in direct economic losses. In 2010, the frequent rainstorms and heavy rainfall in China resulted in floods along many river basins, causing grave secondary disasters such as landslides and debris flows. From May to July, 14 rounds of intense rainfall impacted the south of China, causing severe floods. From mid-July to early September, northern China experienced serious rainstorms and floods. Heavy rains in Zhouqu county and Gansu province triggered dramatic torrential floods and debris flows, and debris flows and landslides hit Yunnan, Guizhou, and Sichuan frequently during the flood season. In July 2012, torrential floods in some reaches of the Hai River Basin exceeded the historical record. During the first 20 days of October, unprecedented persistent severe rainfall struck Hainan. Torrential floods and debris flows in medium and small river basins have had a severe impact in recent years, and the problem of urban water-logging has increased.

1 The Flood Season of 2010 brought Serious Floods to the South and the North

From May to July 2010, 14 rounds of heavy rainfall swamped the south of China, with heavy rain mainly concentrated in Jiangxi, Hunan, Guangdong, Fujian, Guangxi, Zhejiang, and Anhui. Precipitation in some areas reached 800–1200 mm, and on Wuyi Mountain in Fujian province, it reached up to 1481.5 mm. Frequent intense rainfalls once in the flood season maintained high water level in rivers, lakes, and reservoirs, resulting in flooding of the main stream in the upper reaches of the Yangtze River, the Ganjiang River, the Xinjiang River, the Fuhe River, and the Qujiang River, leading to serious economic losses and casualties. In mid-June, the five rivers in Jiangxi, Poyang Lake (in Jiangxi Province,) and the Yangtze River's Jiujiang section, affected by heavy rainfall, surpassed the warning water level, causing dramatic record-breaking floods along the Fuhe River, Xinjiang River, and Ganjiang River, and collapsing the Changkai Dyke along the main stream of Fuhe River. From mid-July to early September, northern and western China was hit by 10 rainstorms, causing floods in the Weihe River, the Liaohe River, and the Second Songhua River.

Debris flows and landslides are serious disasters occurring often in summer. From 21:00 on June 27 to 20:00 on June 28, 260.4 mm of precipitation fell in Gangwu, a town in Guanling county of Guizhou province, where drought and intense early rainfall triggered serious landslides that killed 42 people died and 57 people went missing. On July 13, intense rainfall triggered dramatic floods and debris flows in Xiaohe, a town in Qiaojia county of Yunnan province, in which 19 people died and 26 people went missing. In the evening of August 7, brief but heavy regional rain fell in Gannan, which which is known for short but intense rains, resulting in dramatic torrential floods and debris flows in Zhouqu county, killing over 1,700 people died (including those missing). From August 12 to 23, regional rainstorms repeatedly hit Sichuan Province, causing severe geological disasters such as debris flows and landslides in earthquake stricken areas such as Wenchuan, Yingxiu, and Hanyuan.

Due to the lack of investment channels and development potential for improvement to medium and small rivers, and as a result of low flood control standards, low quality dykes for medium and small rivers basin, and water projects in a state of disrepair that are not well maintained, the riverbeds are raised due to silting, narrowing and water way and making it more shallow. Flood and poorly managed excavation further drastically decrease the rivers' ability to variable water levels. Increased demand and over-exploitation of forest resources severely deteriorates vegetation to an increasing degree in mountains, causing land and soil erosion, and sharply decreasing flood prevention capacity and disaster resistance thereby increasing of the risk of rainstorms and floods.

2 Frequent Water-Logging in Urban Areas

Cities are more fragile and sensitive to meteorological disasters than rural areas due to their dense population and social and economic activity centers. For example, on December 7, 2001, a light snow in Beijing caused a massive traffic jam, which was so severe that it is still remembered today. Rainstorm flooding is one of the predominant meteorological disasters cities face. Some large Chinese cities have been frequently hit by rainfalls over the last few years, severely affecting people's daily lives. In the flood season of 2011 when severe urban water-logging occurred in Shenzhen, Hangzhou, Wuhan, Beijing, Changsha, Chengdu, and Nanjing, intense rainfall caused traffic jams, and streets, houses, and automobiles were flooded, leading to several casualties and property damage. Extreme climate events are clearly increasing in frequency and intensity along with global warming, and the local climate and ecological environment in urban areas is changed even more by urbanization's urban heat traps or wet island effects, which raise the magnitude and frequency of climate events. The failure of design standards for city layouts and arrangements meant to adapt to the features and intensity of new meteorological disasters exacerbate the meteorological disaster risk of primary infrastructure systems including power supply, water supply, gas supply, communication, and transportation. Water-logging is more likely to emerge where second rate infrastructure construction took place, particularly urban sewerage systems and large-scale surface sealing. Between July 21 and 22, 2012, heavy rainstorms hit Beijing. The regional average rainfall was 190.3 mm, exceeding the record for the past hundred years. Near 80 people were killed and serious economic losses were suffered.

3 Influence of Tropical Cyclones

From 1990 to 2010, tropical cyclones caused an annual average death toll of 429 individuals and direct economic losses of 30.7 billion RMB in China. Economic losses grew by 5.39 billion RMB and the death toll decreased by 185 people from 2001 to 2010 compared with 1990 to 2000. Large economic losses and high casualties nevertheless took place during several years in this period. In 2005, the direct economic losses reached 81.47 billion RMB, and in 2006, the death toll reached 1522 people, both numbers ranking second in their respective category for the last 20 years.

In 2004, "Rananim" affected over 18 million people, leaving 169 people dead and causing 20 billion RMB in direct economic losses in Zhejiang, Fujian, Jiangxi, Anhui, Hubei, and Henan. In 2006, super typhoon "Saomai" hit Zhejiang, causing 483 casualties and 19.6 billion RMB in direct economic losses. In the same year, an unprecedented severe tropical storm named "Bilis" swept across seven provinces in the south of China, affecting over 31 million people, killing 843 people, and leading to 34.8 billion RMB in direct economic losses. In 2009, typhoon "Morakot" hit Ali Mountain in Taiwan, carrying about 3,139 mm of precipitation which fell at a rate of over 1,000 mm a day for two days straight, causing the most serious floods in the last 50 years in southern Taiwan, and resulting in serious casualties and property losses. Typhoon "Fanapi" in 2010 left 2.225 million people affected and 129 people dead in Guangdong, Guangxi, and Fujian.

The increase of direct economic losses caused by tropical cyclones is closely related to economic development in coastal provinces over the last 20 years, while the decrease of casualties is partly attributed to enhancements to public disaster prevention awareness and the organizational capability of authorities at different levels to combat disasters.

4 Influence of Cold and Freezing Damages

Under the influence of global warming, the frequency of cold spells in winter and cold and freezing disasters have been generally decreasing, but some extreme cold events still occur at an even greater magnitude. As the economy develops, key infrastructure such as electricity grids, communication lines, and energy connections, become more vulnerable to abnormal climate disasters. In February of 2005, fierce freezing disasters hit some areas in Hunan, Hubei, and Guizhou, and the Hunan power grid suffered the most severe freezing damage since 1954. In early 2008, southern China experienced the worst rain and snow disaster in 50 years, and in some areas the worst in a hundred years, which was particularly widespread, very intense, long lasting, and which had a massive impact, resulting in large losses and serious impediments to traffic, energy supply, power transmission, communication facilities, agriculture, and human life, affecting more than 100 million people and leading to over 159 billion RMB in direct economic losses. That reflects the greatest economic loss and largest population affected for similar disasters over the last 50 years. In January 2011, when the nationwide average temperature fell to its lowest level on record, rainy and snowy weather hit southern China three successive times. The lowest temperature days and longest stretches of freezing rain were in Hunan and

Guizhou, where the average temperature, average maximum air temperature, and average minimum air temperature were the lowest since 1961. The rainy and snowy weather affected Hunan and Guizhou most seriously, affecting about 44 million people, damaging 2.5 million hectares of crops, and leading to over 16 billion RMB in direct economic losses across ten provinces including Guizhou, Hunan, Jiangxi, Anhui, Hubei, and Guangxi.

5 Influence of High Temperatures and Heat Waves

Aside from the impact on agricultural production, hot weather and heat waves also has an effect on industrial production because of increases to electrical loads in cities and towns. Long stretches of high temperatures exert negative impacts on human physiology and psychology and may lead to internal thermal stability disorder and greatly harm human health. Diseases such as heatstroke, cardiovascular diseases, cerebral disorders, influenza, and diarrhea occurred frequently during persistent hot weather. In the midsummer of 2003, persistent high temperatures and low rainfall in the south of the Yangtze River in southern China drastically raised the amount of electricity used for cooling, setting the electrical load at successive all-time highs. During the peak power season, all combustion power plants connected to the power grid in eastern China, central China, and southern China operated at full capacity, and 19 provinces took measures to ration power, jeopardizing the power grids in Shanghai, Jiangsu, Zhejiang, and Fujian. In Shanghai, more than 800 enterprises adjusted their holidays to avoid peak power periods, over 300 enterprises took measures to shift off of peak power load, and nearly a thousand enterprises suffered from power shortages. Due to the severe power shortage, Zhejiang province enacted severe power rationing over the largest area. In Henan, Hubei, Hunan, and Jiangxi, the maximum load of the power grid climbed to its all-time record. The number of heatstroke patients received and cured by both the General Hospital of Nanjing Command and the No.1 Affiliated Hospital of Yangtze University exceeded 200 each.[10] In the summer of 2010, persistent high temperatures and humid weather had a grave negative impact on human health and altered daily life, especially for those who work outdoors and in the open air. The hot weather increased cases of sudden illness and the number of hospital calls. Some people in Inner Mongolia, Shandong, Shaanxi, and Chongqing died of heatstroke.

---

10 China Meteorological Administration 中国气象局, Quan guo qi hou ying xiang ping jia 全国气候影响评价 [National Climate Impact Assessment], (Beijing: Zhongguo Qixiang Chubanshe, 2003), 45.

III Prospects for Future Meteorological Disasters

For 100 years, the average annual temperature in China has increased significantly, at a pace slightly higher than that of the rest of the world's. Under various greenhouse gas emissions scenarios, the warming trend in China will continue, as will increases in extreme high temperatures and heat waves. The areas hit by extreme precipitation will change, and the frequency and intensity of meteorological disasters and secondary disasters will increase.

According to climate predictions based on high, medium, and low greenhouse gas emissions scenarios, China's average temperature will increase by 4.2°, 3.7°, or 2.1° in the 21st century. The projected temperature increases are higher north of the Yangtze River than in the south, with the highest projections for northeastern China and northwestern China the lowest projections for southern China. The temperatures will increase more in autumn and winter than in spring and summer, with temperatures in winter experiencing the most remarkable increases. It is unclear how precipitation patterns will change before. Although there are decreasing precipitation trends over certain years, precipitation will increase after 2040. Average precipitation per decade will decrease slightly in southern China and southwestern China during the beginning of the 21st century, whereas northern China has the highest precipitation increases, followed by northeastern China and northwestern China, and the southern China will have smallest increase throughout the century.[11]

For the next 10 years, except for on Qinghai-Tibetan plateau, the average number of high temperature days annually will increase in China: by 15 to 20 days in southern Xinjiang and the central Huanghuai plain, and by 7 to 15 days in most other areas. The average number of light rain days in China (with daily precipitation of 1–10 mm) will increase by 5% to 25% in the north of northwestern China and in some areas of the Huanghuai plain and the Yangtze River basin, but will decrease by 5% to 25% in northeastern China and in some areas in the Qinghai-Tibetan Plateau as well as in the south of northwestern China. The average number of moderate rain days in China (with daily precipitation of 10 to 25 mm) will decrease slightly in northeastern China, in coastal areas in southern China, in the center of the Qinghai-Tibetan plateau, in some areas in northwestern China and Inner Mongolia, but will increase in the east of

11 Xu Chonghai, Luo Yong, and Xu Ying 许崇海、罗勇、徐影, "Quan qiu qi hou mo shi dui zhong guo jiang shui fen bu shi kong te zheng de ping gu he yu gu 全球气候模式对中国降水分布时空特征的评估和预估 [Assessment and Projection of Spatial and Temporal Distribution of Precipitation in China Based on Global Climate Models]," *Qihou bianhua yanjiu jinzhan* 6 (2010): 398–404.

southwestern China, in most areas of northwestern China, and in some areas in western Inner Mongolia. The number of heavy rain days (with daily precipitation >25 mm) will mainly increase by more than 10%, but it will decrease greatly in some areas of the Huanghuai Plain, in central northeastern China, and in western China, with a maximum decrease of more than 50% in some regions.

During the middle of the 21st century (from 2046 to 2065), under the medium emissions scenario, average annual temperature and average minimum temperature will increase by 2.5°, while the maximum temperature will increase by about 2.4°. The maximum temperature and the minimum temperature will continue to rise, increasing high temperature days and heat waves and decreasing cold days. The 20-year extreme low value for minimum temperature in winter will rise by about 2.9° (and the corresponding extreme high for maximum temperature in the summer will rise by about 2.5°), rising faster than average temperatures. The exponential increase of heat waves and warm night will be more evident in northeastern China, northwestern China, and the Qinghai-Tibetan plateau than in the south of the Yangtze River, with a 40 day increase in Xinjiang and Tibet, and a 30~40 day increase in northeastern China. The warm night index will increase in most of western China, especially in southwestern China, while the decrease in cold days will be most pronounced in the Qinghai-Tibetan plateau, followed by northeastern China and coastal regions.

During the middle of the 21st century (from 2046 to 2065), under the medium emissions scenario, both average annual precipitation and average seasonal precipitation will increase, with the regional mean annual precipitation increasing by about 6%. The average value maximum precipitation for the next 20 years is about 40 mm per day, with relatively large and evident increases in southwestern China and central China. The average number successive rain free days annual and successive rain days annual varies somewhat differently across regions and time periods, but all values for maximum precipitation over five days, extreme precipitation values, and precipitation intensity increase in most of China. There will be more successive rain free days in Xinjiang and the south of the Yangtze River, but fewer in northeastern China, northern China and in the vicinity of Qinghai. Successive rain days will not change significantly in northern China, but will decrease in the south of the Yangtze River, and decrease markedly in southwestern China. The number of days with daily precipitation over 10 mm will increase in all regions excluding northwestern China, increasing most in southwestern China and in the Yangtze River basin. Extreme values will increase similarly across the seasons, except that there will be no clear variations in winter south of 28N. Precipitation intensity will increase throughout China, with the south of the Yangtze River experiencing relatively higher increases.

On account of these variations, it is suggested by the preliminary study that under the medium emissions scenario, hot weather and heat waves will be more frequent in the middle of the 21st century, intense rainfall will increase, and regional flood disaster risks will increase. Extreme precipitation events will increase in the Song-Liao River Basin in summer, raising flood disaster risk. In the Huang-Huai-Hai River Basins, although extreme precipitation does not appear to increase or decrease in spring, increased precipitation variation between years is expected, causing intense rainfall every few years. In the southern river basins, periods of continuous precipitation will shorten while the intensity of extreme precipitation will increase, bringing more rainstorms and higher risks of flood, especially for intense summer rainfall.[12]

Under the medium emissions scenario, during the early and middle 21st century, though precipitation will increase, global warming will increase droughts as well in China. The overall drought area and frequency is projected to increase consistently, causing great concerns as larger swaths are exposed to drought. Areas suffering from severe drought and moderate drought will grow before the year of 2030, though they will shrink thereafter, but remain larger than they were in the late 20th century. Areas suffering from mild drought and light drought will decrease. Frequency of droughts in eastern China will increase much more so than in western China. For the next 40 years, the number of years with one to three months of continuous drought will decline in most regions of China, while the number of years with four to six months of drought will increase in most regions of China. Eastern Inner Mongolia to western Jilin and the southwest region including Chongqing, Guizhou, Guangxi, and Qinghai will experience a decrease of about three years of drought lasting four to six months. The number of years with over six months of drought will increase in most areas of China except for the east of northwestern China and some areas of the Sichuan Basin, while central northeastern China, the northwest of northern China, most parts of southwestern China and Qinghai province will experience the largest increases.[13]

---

12 Xu Chonghai, Yong Luo, and Ying Xu, "[Projected changes of precipitation extremes in river basins over China]," *Quaternary International* (2011), doi: 10.1016/j.quaint.2011.01.002.

13 Xu Chonghai, Luo Yong, and Xu Ying 许崇海、罗勇、徐影, "IPCC AR4 duo mo shi dui zhong guo di qu gan han bian hua de mo ni ji yu gu IPCC AR4 多模式对中国地区干旱变化的模拟及预估 [Simulation and Prediction of the Drought Variations in China by IPCC AR4 Multi-Model Ensemble]," *Bingchuan dongtu* 32 (2011): 867–874.

Based on the temperature and precipitation trends and the extreme climate events index projection for the next 10–50 years in China, there will be more meteorological disasters and secondary disasters in China in the future:

A *Regional Flooding*

Since intense rainfall (the precipitation heavier than heavy rainfall) will increase, there will be more regional floods in China in the next ten years. Due to the great variation in the spatial distribution of extreme precipitation, the area suffering from regional floods is also expanding. The probability of regional flooding is relatively higher in such areas as the south of northeastern China, northern China, the Huanghuai Plain, the east of northwestern China, northern Xinjiang, and the east of southwestern China.

B *Drought*

In the next 10–50 years, periodic droughts and successive seasonal droughts are more likely to happen in China, as the frequency of regional drought rises and the area covered by droughts expands. There is a relatively higher probability of regional droughts in the east of southwestern China, southern China, the north of northern China, and eastern Inner Mongolia.

C *High Temperatures and Heat Waves*

As temperatures rise, there will be more heat waves in China and a higher probability of extreme hot weather and heat waves in Xinjiang, northern China, the Huanghuai Plain, the middle and lower reaches of the Yangtze River, as well as the southern region of the Yangtze River. These phenomena may cause further serious social, energy, and water resource problems. The rising temperature will also raise forest and prairie fire danger ratings.

D *Debris Flow*

In the next ten years, increasing regional floods will cause more frequent debris flows, especially in areas such as the east of northwestern China and the east of southwestern China, which are more vulnerable to geological disasters.

E *Prairie and Forest Fire Danger*

In the next ten years, global warming and increasing droughts will raise forest and prairie fire danger in eastern Inner Mongolia, the west of northeastern China, the east of southwestern China, the southern region of the Yangtze River, and southern China.

Given the prospect of increasing meteorological disasters, China should strengthen weather and climate change monitoring efforts, invest in research on factors and mechanisms that can reduce meteorological disasters, and on the social vulnerability and sensitivity to the meteorological disasters, and enhance the meteorological disaster risk assessment and early warning systems. It is necessary to establish and improve a mechanism to defend against meteorological disasters, which should be led by the government, integrated among relevant departments, and include public participation. China should further strengthen climate change impact assessments for sensitive sectors, major projects, and city planning, as well as improve meteorological disaster risk management and enhance the public's capability of avoiding meteorological disaster risk, in order to adapt to climate change.

(This article was originally published in Chinese in 2012.)

CHAPTER 15

Extreme Climate Projections and Risks in China

Ying Xu, Chonghai Xu and Jing Feng

Abstract

Simulations from five global climate models from the fifth phase of the Coupled Mode Inter-comparison Project (CMIP5) organized by IPCC AR5 are used to analyze changes in extreme climate in China for the future. Results indicate that with new greenhouse gas emissions China is experiencing more heat waves, warm nights, cold days and a greater frequency, strength, geographical spread, and duration of some extreme precipitation indexes. The additive effect of these changes could lead to unprecedented extreme weather and climate events. The possibility of an ever greater number of people and facilities exposed to extreme climate are increasing, as is the risk of disaster.

Keywords

RCPs emission scenarios – Projection of extreme climate events – climate change adaptation

I Introduction

In 2012, the special report on "Managing the Risks of Extreme Events and Disasters to Advance Climate Change Adaptation" (SREX) was published in joint coordination with Working Groups I (WGI) and II (WGII) of the Intergovernmental Panel on Climate Change (IPCC). It touches on various topics, but focuses on how adaptation and disaster risk management for extreme events. The report points out that changing climate could lead to extreme

* Ying Xu, National Climate Center, China Meteorological Administration, Ph.D. Research areas: climate change detection, climate change projection, climate disaster risk assessment. Chonghai Xu, Meteorological Observation Center, China Meteorological Administration, Ph.D. Research areas: climate change projections. Jing Feng, master. Research areas: projection of extreme weather events.

 | DOI 10.1163/9789004274648_016

weather and changes in the frequency, intensity, geographical and seasonal appearance, duration, and scope of climatic events, and result in unprecedented extreme weather and climate events.[1]

The collective results of the report indicate that since 1950 the observed trends of some extreme events had changed. There is a high likelihood that the number of cold days and nights decreased, and the number of warm days and nights increased. There was a warming trend in daily extreme temperature and the number of heat waves increased. Statistically significant trends were noticed in the number of heavy precipitation events in some regions. Some regions of the world have experienced and longer more intense droughts, and the magnitude and frequency of floods have changed across different regions. The economic loss resulting from extreme weather and climate events has also increased significantly. With an increasing concentration of different greenhouse gases, the frequency and magnitude of extremely warm events will grow around the world, as extremely cold events decrease. The duration, frequency, and intensity of heat waves will also increase. The frequency of strong rainfall or the proportion of total rainfall falling as heavy participation may increase. The increase of heavy rain will lead to more local floods in some river basins and regions. Heat waves, glacier retreat, and permafrost degradation will increase flood landslides.

Since 1951, the frequency and intensity of extreme weather and climate events in China have similarly changed in different patters across China's various regions. The maximum and minimum temperatures, heavy precipitation, droughts, and typhoons have all been altered. Heavy precipitation events have increased and grown in the middle and lower reaches of the Yangtze River, and in South-Eastern and Western regions; precipitation frequency has decreased significantly across the country; the area suffering from droughts has grown, especially in northern China and northeastern China; an unmistakable decrease in cold nights, cold days, cold waves, and the number of frost days was seen across the country; whereas an increase in the number of warm nights and warm days was also witnessed; finally, the number of typhoon landings decreased. How will extreme climate events in China change as more greenhouse gasses are present in the atmosphere? How does China adapt to extreme climate changes and and adopt disaster risk management? In this article, the five global models' simulations using various greenhouse gas emissions scenarios are used to analyze the possible future changes to extreme climate events in China, and some adaptive management measures are proposed.

---

1 IPCC, 2011, "Managing the Risks of Extreme Events and Disasters to Advance Climate Change Adaptation," *Special Report of the Intergovernmental Panel on Climate Change.*

II RCPs Emissions Scenarios and Data

A *RCPs Emissions Scenarios*

Before the IPCC AR5, the IPCC special report offered six kinds of common emissions scenarios and the corresponding warming estimates. Each emissions scenario focuses on regional development, but is unable to fully reflect the UNFCCC goal of stabilizing atmospheric concentrations of greenhouse gases. In the IPCC AR5 for global climate change projections, new greenhouse gas emissions scenarios make use of RCPs (Representative Concentration Pathways), which regard unit area radiation force to represent future emissions scenarios. Compared with the prior emissions scenarios, RCPs offer a wider range, more objective projections, and are more convenient for use in areas such as agriculture and water resources and in climate change assessment research.[2]

In the RCPs emissions scenarios, four different paths are distinguished by total radiative forcing: a higher emissions path, where radiation force reaches 8.5 W/m^2 by 2100, and continues to rise; two stabilization paths, where the radiative forcing will roughly stabilize at 6 W/m^2 or 4.5 W/m^2 after 2100; and a lower emissions path, where radiative forcing level first reaches a value around 3.1 W/m^2 in the middle of 21st century, and the returns to 2.6 W/m^2 by 2100. These emissions scenarios include the emissions and concentration time paths of all the greenhouse gases, aerosols, chemically active gases, and land use, and further project into the short term, until 2035, and into the long term, until 2300, to meet the needs of different subjects and research groups. The characteristics of different RCPs emissions scenarios are shown in Table 15.1.

TABLE 15.1 *Characteristics of concentration paths*

Name	Radiative Force	Equivalent concentration
RCP 8.5	8.5 W/m^{-2} in 2100	≈1370 CO_2–eq
RCP 6.0	Radiative force stabilizes at 6 W/m^{-2} after 2100	≈860 CO_2–eq
RCP 4.5	Radiative force stabilizes at 4.5 W/m^{-2} after 2100	≈650 CO_2–eq
RCP 2.6	Radiative force less than 2.6 W/m^{-2} after 2100	≈490 CO_2–eq

Data source: Lin Erda, Liu Yingjie. "Advance in New Scenarios of Greenhouse Gas Emission and Climate Change," *Scientia Agricultura Sinica*, 2008, 41 (6): 1700–1707.

2 Lin Erda, Liu Yingjie, "Advance in New Scenarios of Greenhouse Gas Emission and Climate Change," *Scientia Agricultura Sinica*, 2008, 41 (6): 1700–1707.

B Data and Definitions

1 Data Sources

Data used in the study comes from the five global climate model results from CMIP5. CMIP5 has more multiple model simulations than CMIP3, such as long-term historical climate simulations (historical), natural force simulations, greenhouse gas forced simulations, as well as other forced simulations, which are better suited for climate change detection and attribution studies. Experiments predicting future climate change include simulations forced by RCP 2.6, RCP 4.5, RCP 6.0 and RCP 8.5.

The global climate models in CMIP5 are atmosphere-land surface-ocean-sea ice coupled climate system models, which consider changes in solar radiation, greenhouse gases, and sulphate aerosols. Some models also consider volcanic activity's effect on aerosol. Unlike in CMIP3, all models in CMIP5 include global carbon cycle processes and dynamic vegetation processes. The greatest improvements over CMIP3 are the increase of horizontal resolution of the atmosphere and ocean models, dynamic frameworks for atmospheric circulation, introduction of new radiation schemes, improvements in the treatment program for aerosol, flux in tropospheric and stratospheric representations, particularly of retaining water vapor in the atmosphere model, and improvements to couplers to avoid energy loss and calculation drift. These improvements bring the simulation results closer to real historical changes of the climate (see http://pcmdi-cmip.llnl.gov/cmip5/availability.html, more information on the model can be found).

This study uses the simulations from the five global climate models from CMIP5. The changes of extreme climate events under RCP 4.5 and RCP 8.5 for 2006–2099 are analyzed in Table 15.2. All of the data are interpolated onto a common 1°×1° grid.

TABLE 15.2 *Information for the five global climate models*

Model Name	Institution	Atmosphere model
BCC-CSM1–1	BCC, China	64×128/T42L26
CanESM2	CCCma, Canada	64×128/T42L35
CNRM-CM5	CNRM and CERFACS, France	128×256/T127L31
INMCM4	INM, Russia	120×180
NorESM1–M	Norwegian Climate Centre, Norway	96×144/f19L26

Source: http://pcmdi-cmip.llnl.gov/cmip5/availability.html.

2 Definition of Extreme Climate Events

Studies of extreme climate events are carried out by analyzing climate extremes indexes as defined by Frich et al. (2002).[3] These include a set of extreme temperature indices: HWDI, TN90, and TX10, and a set of extreme precipitation indices: CDD, CWD, R10, R5D, SDII, and R95T (more information in Table 15.3).

TABLE 15.3 *Definition of extreme temperature and precipitation indices*

Name	definition	units
HWDI	Heat wave duration index: maximum period > 5 consecutive days with Tmax > 5 deg C above the 1961–1990 daily Tmax normal	days
TN90	Warm nights. Fraction (expressed as a percentage) of time Tmin > 90th percentile of daily minimum temperature, where percentiles are based on the 1961–1990 base period	%
TX10	Cold days. Fraction (expressed as a percentage) of time Tmax < 10th percentile of daily minimum temperature, where percentiles are based on the 1961–1990 base period	%
CDD	Maximum number of consecutive dry days (Rday < 1 mm)	days
CWD	Maximum number of consecutive wet days (Rdays ≥ 1mm)	days
R10	No. of days with precipitation greater than or equal to 10 mm/day	days
R5D	Maximum 5 day precipitation total	mm
SDII	precipitation intensity. Simple daily intensity index: annual total/number of Rday greater than or equal to 1 mm/d	mm/day
R95T	Fraction (expressed as a percentage) of annual total precipitation due to events exceeding the 1961–1990 95th percentile	%

3 Frich P., et al. 2002, "Observed coherent changes in climatic extremes during the second half of the twentieth century," *Clim. Res.*, 19: 193–212.

III Projections of Extreme Climate Change in China under the RCPs

A *Projected Changes to Extreme Temperature*

Table 15.4 shows the projected changes for some extreme climate indices related to temperature based on results from the five global climate model simulations. For the heat wave duration index (HWDI), most models show an increasing trend under the RCP 4.5 and RCP 8.5, except for the INMCM4 which shows a decreasing trend for RCP 4.5. The trend is 0.08~11.05 days/10a under RCP 4.5 and 1.82~18.1 days/10a under RCP 8.5. The linear trend under RCP 8.5 is clearly higher than under the RCP 4.5 scenarios, by a range of 6~9 days/10a.

Warm nights (TN90) also display positive linear tendencies, 0.24~3.17%/10a under RCP 4.5 and 2.0~5.33%/10a under RCP 8.5. The trend is significantly higher under RCP 8.5 than under RCP 4.5, and the difference between the linear regressions is in the range of 2~4%/10a.

Cold day events (TX10) using most models display a decreasing trend under the RCP 4.5 and RCP 8.5, except for the CanESM2 model which presents a weak increasing trend under the RCP 4.5. The trend is –1.79~0.17%/10a under RCP 4.5 and –2.19~0.4%/10a under RCP 8.5. The decreasing trend under RCP 8.5 is far stronger than under the RCP 4.5 scenarios.

TABLE 15.4 *The projected linear regressions for some extreme climate indices related to temperature from 2006 to 2099 (relative to the 1986–2005 years)*

	HWDI (days/10a)		TN90 (%/10a)		TX10 (%/10a)	
	RCP4.5	RCP8.5	RCP4.5	RCP8.5	RCP4.5	RCP8.5
BCC-CSM1–1	3.06	12.22	1.86	5.11	–0.44	–0.68
CanESM2	11.05	18.1	3.17	5.33	0.17	–0.4
CNRM-CM5	2.72	8.31	2.12	4.76	–0.53	–0.78
INMCM4	–4.06	1.82	0.24	2.0	–1.79	–2.19
NorESM1–M	0.08	9.51	1.45	4.2	–1.59	–1.84
Ensemble	2.57	9.99	1.77	4.28	–0.83	–1.18

Source: Jing Feng, 2012. "Simulation and Projection for Regional Climate in China by Multiple Global Climate Models." M.S.'s thesis, Nanjing University of Information Science & Technology

Under the RCP 4.5 and RCP 8.5 scenarios, the HWDI are always increasing from the benchmark average of 1986–2005. The increases for HWDI range between 10% to 20% through the end of the 21st century. For TN90, the increases range from 4% to 7%. For TX10, the declines range from 60% to 80%.

The geographic distribution of projected extreme temperature indices' linear trends demonstrates that HWDI will increase across the entire country under RCP 4.5 scenarios, but especially in the northern parts of northeastern and northern China, Xinjiang, and the southern parts of Tibet. Under RCP 8.5, the rate of increase is higher than under RCP 4.5, but the geographic distribution remains consistent.

The geographic distribution of the extreme warm nights is less pronounced in northern parts of China than in the south, and the most pronounced projected increases are found in southwestern China, southern China, the eastern coast and the Tarim basin. The linear trend for RCP 8.5 is unsurprisingly higher than for RCP 4.5 overall, but the increasing trends in the Tarim basin are equivalent to the trends for southwestern and southern China.

Cold days under RCP 4.5 and RCP 8.5 scenarios both display a downward trend, especially in northeastern China and the in Qinghai-Tibet plateau.

B *Projected Extreme Precipitation Changes*

Table 5 shows the linear trends of CDD under the RCP 4.5is –0.6~–0.16days/10a and –1.19~–0.07days/10a under RCP 8.5. For the end of the 21st century (2080–2099), the CDD change reaches –8% for RCP 4.5 and –11% for RCP 8.5.

The linear trend for CWD under the RCP 4.5 is 0.15~0.49%/10a and 0.07~0.57%/10a for RCP 8.5. For the end of the 21st century (2080–2099), the CWD change reaches 5% for RCP 4.5 and 8% for RCP 8.5.

R10 reveals a linear trend of 0.24~0.56%/10a for RCP 4.5 and 0.36~0.74%/10a for RCP 8.5. For the end of the 21st century, the changes reach 14% for RCP 4.5 and 20% for RCP 8.5.

The linear trend for R95T is 0.1~0.71day/10a for RCP 4.5 and 0.42~1.07days/10a for RCP 8.5. For R5D, the trend is 0.1~1.88%/10a for RCP 4.5 and 0.56~3.61%/10a for RCP 8.5. For SDII the trend is 0.02~0.08%/10a for RCP 4.5 and 0.04~0.11%/10a for RCP 8.5.

All extreme precipitation indices trend upwards, except CDD. It highly likely that there will be more extreme precipitation events and that they will also be stronger and more intense in the future.

CDD is projected to decrease in most parts of China under RCP 4.5 and RCP 8.5, specifically in Xinjiang. CWD is projected to increase in most parts of China, expect in the Qinghai-Tibet plateau where it shows a decreasing trend. R10 has an increasing trend in most parts of China, with the highest values under RCP 4.5 in the southern and eastern edge of the Tibetan plateau, the Qinling Mountains, Guangdong, and Guangxi, and under RCP 8.5 in the southern parts of the Qinghai-Tibet plateau.

TABLE 15.5 *The projected linear trends for some extreme climate indices related to precipitation from 2006 to 2099 (benchmarked at 1986–2005)*

	CDD (days/10a)		CWD (days/10a)		R10 (days/10a)	
	RCP 4.5	RCP 8.5	RCP 4.5	RCP 8.5	RCP 4.5	RCP 8.5
BCC-CSM1–1	–0.16	–0.12	–0.11	–0.3	0.24	0.36
CanESM2	–0.56	–0.19	0.19	0.07	0.46	0.60
CNRM-CM5	–0.60	–0.80	0.49	0.57	0.26	0.40
INMCM4	–0.19	–0.07	0.26	0.22	0.25	0.45
NorESM1–M	0.09	–0.26	0.15	–0.25	0.56	0.74
Ensemble	–0.29	–0.49	0.20	0.06	0.35	0.51

	R95T (%10a)		R5D (mm/10a)		SDII (mmday–1/10a)	
	RCP 4.5	RCP 8.5	RCP 4.5	RCP 8.5	RCP 4.5	RCP 8.5
BCC-CSM1–1	0.39	0.94	0.93	2.54	0.04	0.09
CanESM2	0.50	1.07	1.49	3.61	0.06	0.10
CNRM-CM5	0.11	0.59	0.74	1.80	0.03	0.07
INMCM4	0.10	0.42	0.10	0.56	0.02	0.04
NorESM1–M	0.71	1.02	1.88	2.76	0.08	0.11
Ensemble	0.36	0.81	1.03	2.25	0.04	0.08

Source: Jing Feng, 2012. Simulation and Projection for Regional Climate in China by Multiple Global Climate Models. M.S.'s thesis, Nanjing University of Information Science & Technology

R95T has a weak decreasing trend in the northwestern part of the Hosi Corridor, but it increases in most other area of China under RCP 4.5. Under RCP 8.5, high value areas are found mainly in the Qinghai-Tibet plateau and in Yunnan Province. The highest value of SDII is located in southern parts of the Tibetan plateau.[4]

4 Jing Feng, 2012, "Simulation and Projection for Regional Climate in China by Multiple Global Climate Models." M.S.'s thesis, Nanjing University of Information Science & Technology.

IV The Possible Risks from Changes to Extreme Climate Events

Warm events under the RCP 4.5 and RCP 8.5 scenarios, such as heat waves and warm nights, are projected to increase. Cold day events will become scarcer, although the increase in warm events is stronger than the decrease in cold event. The linear trend shifts under are more obvious under high emissions scenarios than under medium emissions scenarios. By the end of 21st century, HWDI grow 7.3 times greater under RCP 4.5 and 18 times greater under RCP 8.5, warm nights will become 3.8 times more frequent under RCP 4.5 and 7.2 times more frequent under RCP 8.5, and cold days will increase by 60% under RCP 4.5 and by 80% under RCP 8.5. There is a strong possibility of more extreme precipitation events that are stronger and more intense in the future. All extreme precipitation indices increase, except for CDD. By the end of 21st century, the following changes are predicted under RCP 4.5: CDD –8%, CWD +8%, R10 +14%, R95T +18%, R5D +15%, SDII +8%. The changes predicted for RCP 8.5 are: CDD –11%, CWD +5%, R10 +20%, R95T +38%, R5D +27%, SDII +13%.

The projected results indicate that under the greenhouse gas emissions scenarios, the changing climate can lead to significant changes of frequency, intensity, spacial distribution, and duration of extreme weather and climate events in China and across the globe, resulting in unprecedented extreme weather and climate events. A greater number of people and facilities will likely be exposed to extreme climates, and the risk of disaster will continue to rise. The increase in climate disasters will exacerbate income inequality between poor and rich areas around the world. Changes to the spatial distribution, the intensity, and frequency of climate disasters shifts them to new areas and regions, which causes serious losses in areas suffering from poverty, and further deteriorates their disaster recovery capability.

Extreme climate, as one of the factors affecting disaster risk, will also have a greater impact on different climate change related industries (such as water conservation, agriculture, food security, forestry, health, and tourism). The effect of extreme weather events will greatly depends on the degree of exposure and the vulnerability of specific areas. For example, if a disaster occurs in an area with a greater concentration of exposed population and economic facilities, the loss from the weather and climate disaster will increase. Urbanization and changes to social and economic conditions have an influence on the degree of vulnerability and exposure to extreme climate as well. For example, growing urban residential areas in coastal regions affect the ability of natural coastal ecosystem to effectively adapt to the extreme climate events, making the regions more vulnerable. More frequent floods will destroy the cities and food production, lower food security, and deteriorate conditions in poor areas. Heavy rains and floods can also contaminate surface water and damage the health of the urban environmental. Although a great deal of uncertainty exists in estimating changes to small watersheds, it is certain that climatic changes will have a serious impact on water management systems.

New challenges are posed by the changing extreme climate events, threatening national disaster risk management, and requiring changes to adaptability, responsiveness, flexibility in the face of these new risks. Risk sharing and transfer mechanisms should be established across different levels, on the local and national scale, and new financing means should be provided to improve the adaptability to extreme events.

(This article was originally published in Chinese in 2012.)

CHAPTER 16

Coastal Cities' Adaptation to Climate Change

Cao Lige, Su Buda, Zhai Jianqing,
Marco Gemmer and Zhan Mingjin

Abstract

The economically, socially, and culturally most developed areas worldwide are located predominantly along the coastlines. Socioeconomic development has accumulated large populations and high value assets in coastal cities while coastal cities have become prone to climate change effects. Sea level rise, tropical cyclones, rainstorms, floods, storm surges, high temperatures, and heat waves have posed new challenges to coastal city development. Domestic and foreign coastal cities such as Shanghai, Guangzhou, Tianjin, Xiamen, Hong Kong, New York, London, Tokyo, Sydney, Venice, Singapore, and Cape Town are analyzed on their current climate change adaptation techniques. Measures are compared and future challenges for adaptation are assessed.

Keywords

coastal cities – adjustment – climate change

Coastal areas boast the highest population density in the world. More than half of the populations worldwide live within a 100 km belt along coasts. This proportion is projected to increase by 25% over the next 20 years. The majority

* Cao Lige, National Climate Center of the China Meteorological Administration, master degree candidate, assistant researcher, research field: climate change and risk management; Su Buda, National Climate Center of the China Meteorological Administration, post-doctoral and associate professor, research field: climate change impact assessment, climate risk management, meteorological index insurance; Zhai Jianqing, National Climate Center of the China Meteorological Administration, post-doctoral, associate professor, research field: climate change impact assessment, climate change impact on water resource and flood response; Marco Gemmer, National Climate Center of China Meteorological Administration, professor, research field: climate change impact assessment; Zhan Mingjin, National Climate Center of China Meteorological Administration, Ph.D student, research field: climate change impact assessment.

 | DOI 10.1163/9789004274648_017

of global economic assets originate from these areas.[1] Coastal areas have the most highly developed economy, social structure, culture, and population density. Ports and cities along the coasts have supported economic prosperity and development and have acted as a critical driving force, promoting local urbanization. Resources and infrastructure are concentrated along the coasts and the most adverse impacts of climate change will most likely occur in coastal areas as well. Dense urban populations have increased the vulnerability to climate change. The probability of typhoons has increased due to sea level rise, as have the chances of rainstorms, floods, storm surges, high temperatures, heat waves and other weather disasters. It is of vital importance for governments in coastal areas to adopt practical and effective measures for climate change adaptation.[2] Research shows that the costs of adaptation are huge but vary according to the scale of sea level rise. Coastal city planners and governments shall prioritize strengthening climate change adaptation measures.

I Climate Change Risk Analysis for Coastal Areas

A *Risk of Sea Level Rise*

Sea level rise is a long-term oceanic phenomenon with additive effects. It will intensify storm surges, coastal erosion, seawater intrusion, soil salinization, and salt tides. The most direct influence is seawater intrusion, which increases the salinity of urban ground water, the primary water source for coastal cities. Rising sea level also raises tide levels, posing a severe threat to the natural urban environment, the safety of drinking water, industrial production, and ecological systems. The habitat of species living and breeding in fresh water before entering the sea is threatened and some have already gone extinct. Aside from the increased strength of typhoons, higher coastal waves pose threats to maritime dykes and to soil in case of dyke failure. Sea level rise affects flood control facilities and drainage systems, the city's self-cleaning capabilities, and thus enhances the probability of urban water-logging. Shipping passages will be minimized due to lower passage ways under bridges. Major typhoon events are likely to become more frequent, so that what was once a once in a millennium event will become a once in a century event.

---

1 Xu Y.J., Singh V.P. (Eds), *Coastal environment and water quality*, (Highlands Ranch, USA: Water Resources Publications, LLC, 2006), 1–3.

2 Duan Xiaofeng, Xu Xuegong, "Research progress and perspective of risk assessment of sea level rise," *Transactions of Oceanology and Limnology*, 2008, (4): 116–122.

B Urbanization and Climate Risk

Ports and cities have supported economic development along coastlines and will further drive urbanization. Climate change and urbanization are two major factors which increase risk and vulnerability. These two factors overlap in coastal cities, which make these areas susceptible to disasters and economic losses.[3] Concentrated urban population offer more opportunities in the service sector, but are susceptible to risks of greater losses from natural and urban disasters and from climate change. The unique structure of cities, including an economy centered society, makes the city vulnerable to climate risks. A rainstorm can cause a series of secondary interruptions such as water-logging and traffic interruptions. Dense fog can lead to expressway closures, air traffic interruptions, and electrical transmission failures causing massive blackouts, thereby directly threatening the normal city life. A city depends upon basic infrastructure such as traffic, power, communication, water, gas supply, and sewage disposal. If a meteorological disaster causes damage to basic infrastructure, the disaster will impact the entire city. The impact of climate change on the environment can lead new invasive species and pests.

C Influence of Extreme Climate Events upon Coastal Cities

Climate change has the potential to increase the frequency and negative economic impact of extreme events. These include sea level rise, storm surges, typhoons, rainstorms, floods, high temperatures, heat waves, thunder, and lightning. Stronger typhoons have posed a severe threat to urban residents and their property, the urban economy, and traffic in coastal cities. More frequent and stronger thunder and lightning storms, which can cause floods, road damage, traffic congestion, blackouts, or damage to electronic information systems, have a detrimental effect on the normal function of society, the economy, and on infrastructure in the city. Urban floods have also become a serious challenge. Dense fog and droughts have challenged disaster prevention systems and confronted fragile infrastructure and public security emergency systems in many cities. Urban traffic is highly sensitive to weather and climate, so that disasters easily increase the cost of traffic infrastructure construction, reduce operation efficiency, and increase potential safety hazards.

D Influence of Comprehensive Disaster Risk in Coastal Cities

Increasing populations, buildings, production, and wealth has in coastal cities has increased disaster risk. Urbanization and modernization have increased

3 The World Bank, *Climate Change Adaptive City Entry Guide*, (Beijing: China Financial Publishing House, 2009), 2–3, 17.

the risk, the frequency, and the impact of disasters.[4] Urban disaster risk has many new features and developmental tendencies: a) disasters have progressed from single, isolated events to common phenomena; b) disasters have become increasingly frequent; c) disasters have begun incorporating multiple complex factors; d) local disasters can spread to become an urban crisis; and e) a crisis in one country could transform into a transnational crisis.[5] Development and spatial management in coastal cities requires harmonization with disaster risk management and climate change projections, which are now being integrated with urban development planning and management.

II Climatic Risk Management in Coastal Cities in China

Coastal areas in China have dense populations and high economic output. The majority of large cities are located in eight coastal provinces, one autonomous region, two provincial municipalities, Hong Kong, Macau, and Taiwan. There are more than 20 cities at or above a medium size, and in average, there is one coastal city every 720km.[6] Smaller cities along the coast are rare. Counting cities that have reached a medium development level or higher, there are about 500 cities and ports of different sizes along the 18,000 km coastline of China, forming an urban belt with an advanced economy, society, and culture. The terrain in most of these areas is low-lying and flat, leaving it vulnerable to various ocean disaster threats caused by sea level rise. Problems such as climate change and sea level rise have affected regional sustainable development.

The river delta areas along China's coastlines have the highest degree of urbanization and the most concentrated population and wealth. Sea levels have risen for the last 30 years. Coastal cities in China such as Shanghai in the Yangtze River Delta, Guangzhou in the Pearl River Delta, Tianjin on the Bohai Sea, Hong Kong, and Xiamen, have experienced different impacts from climate

4 Chen Jing, Liu Jing, Wang Zhiqiang, et al., "Issues and Strategies for Integrated Urban Disaster Risk Management in China," *Journal of Natural Disasters*, 2006, 15 (6): 17–22.

5 Zhang Jiquan, Zhang Hui, Okada Norio, "Integrated Urban Disaster Risk Management: An Innovative Approach and Challenge in the 21st Century," *Human Geography*, 2007, 22 (5): 19–23.

6 Including Dandong, Dalian, Huludao of Liaoning Province; Qinhuangdao of Hebei Province; Tianjin; Yingkou, Longkou, Yantai, Weihai, Qingdao of Shandong Province; Lianyungang of Jiangsu Province; Shanghai; Zhoushan, Ningbo, Taizhou, Wenzhou of Zhejiang Province; Fuzhou, Quanzhou, Xiamen of Fujian Province; Shantou, Shanwei, Guangzhou, Shenzhen, Zhuhai, Zhanjiang of Guangdong Province; Sanya and Haikou of Hainan Province; Beihai, Qinzhou, Fangcheng of Guangxi Province; and Hongkong, Macao and some cities of Taiwan.

change and sea level rise depending upon their local governments, which have taken targeted adaptation measures according as spelled out in their development plans (See Table 16.1). Cities have taken emergency measures along with their long-term plans according to their local needs and specific weather phenomena. Guangdong Province is a large province along the southern coast with nearly 95 million people. It has developed very rapidly and the salt tides have seriously influenced its urban development. In 2005, the most severe salt tide in the past 20 years occurred in the Pearl River Delta, damaging industrial and agricultural production as well as daily life. To cope with the disaster, the first large-scale, long distance, and trans-provincial water transfer was implemented to supply fresh water. In April 2010, the sea level in the coastal area of Guangdong was above average, and water resources in the upper reaches of the Pearl River were low due to the drought in southwestern China. The Pearl River estuary seriously suffered from salt tide intrusion a second time. Guangdong Province took many measures, including water supply planning and construction of water supply systems in the Pearl River Delta, to ensure normal water supply in the Pearl River Delta, Macau, and in other areas during salt tides.

TABLE 16.1 *Climate change risks and adaptation measures in development plans of selected coastal cities in China*

City	Climate change impacts	Adaptation measures
Shanghai (Yangtze River Delta)	Flood disaster and seawater intrusion caused by typhoon, storm surge and river flooding	1. Strengthen sea level rise monitoring and establish an early warning system; 2. Strengthen urban constructions and infrastructure to protect water quality at the source; 3. Strengthen scientific research and establish estuary function zones; 4. Improve regional meteorological services and develop comprehensive measures to address climate change.
Guangzhou (Pearl River Delta)	Extreme events such as sea level rise, typhoons and flooding. The strong typhoons and severe storm surges cause sea-water intrusion. The warming	1. Develop macro strategies to address climate change. During the 11th Five Year Plan, the government spent 158 billion RMB in disaster prevention and mitigation, disaster monitoring, water resource management, and marine

City	Climate change impacts	Adaptation measures
	climate increases haze and high temperature days. Water and electricity become insufficient.	environment, wetland, and mangrove protection; 2. Strengthen infrastructure construction, improve resilience of agriculture and coastal area; 3. Improve weather monitoring and forecasting. Establish climate information systems and strengthen climate change education to raise public awareness.
Tianjin (Bohai Sea)	The Tianjin Binhai New Area is heavily affected by rising sea levels and storm surges. With sea level rise, dykes are becoming less effective. Various tide levels have varying impacts on Tianjin's coastal area. As the sea rises less than half of the dyke facilities are able to protect against the historical high tide level or the once in a century maximum tide level.	1. Make administrative decisions regarding the vulnerability to sea level rise; 2. Strengthen the impact assessment of sea level rise; 3. Control the exploitation of ground water and improve water conservation facilities; 4. Raise dyke design standards and strengthen the construction, maintenance, and supervision of protection facilities; 5. Restore and reconstruct the ecosystem of the region and protect coastal ecological resources of the Binhai New Area.
Hong Kong SAR	Hong Kong, the largest port in Asia and the most important financial city, is located in the Pearl River delta. It is threatened by sea level rise, storm surges, sea-water intrusion and other kinds of extreme weather.	The Hong Kong government has adopted active climate change mitigation strategies so that the energy intensity in 2030 will be reduced by 25% compared with 2005 levels. The main mitigation strategies include fossil fuel reduction, using renewable energy, promoting green buildings, and implementing public energy saving measures. Water saving measures such as sea water utilization, fresh water conservation and water dispatching from Dongjiang River are also included.

TABLE 16.1 (*cont.*)

City	Climate change impacts	Adaptation measures
Xiamen	Xiamen is located in the coastal area of Fujian Province near the Taiwan Strait. In the last 50 years, the average temperature, the extreme high temperature and the number of storm days have all increased. The average wind speed and relative humidity are decreasing. Haze days and fog days are increasing, reducing visibility on shipping routes.	Typhoons, storms, heat waves, and city water-logging caused by extreme precipitation or typhoons have caused large economic losses in Xiamen. The Xiamen government has raised flood control standards to deal with storms expected once every 50 years instead of once every 20 years. The government has strengthened sea level rise monitoring and marine and coastal zone environment management. In order to create a good environment and protect endangered species, the government has constructed artificial beaches and expanded mangrove planting areas.

Data sources: climate data web sites, city climate action reports, "Climate Change Adaptive City Entry Guide," "Global Climate Change and Vulnerability Assessment of Estuary City—A case Study of Shanghai."

III Progress of Low-Carbon Development and Adaptation Actions in Cities Overseas

"Low-carbon development" has become a common target around the world. Many countries and regions have emphasized climate change mitigation efforts such as carbon emissions reduction, but adaptation to increasingly frequent and intense meteorological disasters has not gained much attention. After experiencing numerous extreme weather events following the Cancun Conference, governments have begun discussing research on the impacts of climate change and adaptation possibilities.

A *Climate Change Mitigation and Adaptation in Cities in China and Abroad*

Coastal cities can adjust measures and policies aimed at low-carbon development and climate change adaptation (See Table 16.2). When they develop plans for green and circular economies or for ecological and livable city construction, they can simultaneously incorporate urban risk management and disaster prevention capabilities. Coastal areas can take active climate change adaptation measures such as wetland protection, estuary and alluvial plains development along the coast, coastal erosion control projects, and stronger disaster warning and standards. The "Urban Climate Protection in Dusseldorf and Wuxi" project sponsored by Stiftung Mercator (Germany) has focused on the adverse impacts of climate change and explored measures and means to adapt to climate change, aside from climate protection and energy conservation strategies, by reshaping traffic, industry, real estate development, and urban planning.

Two of the best known tourist cities in the world, Cape Town and Sydney, have focused municipal attention on the combination of environmental protection and climate change adaptation measures in local industry development planning. Sydney launched the first "Earth Hour"[7] in 2007, drawing more than 2.2 million families and enterprises to participate.

TABLE 16.2 *Climate change risks and adaptation measures in development plans of foreign coastal cities*

City	Climate change impacts	Adaptation measures	Specific practices
USA (New York)	New York is not only the biggest city and largest port in the US, but also the most important commercial and financial center of the world. Because of the heat island effect, the annual average temperature in New York has risen more quickly	The climate change adaption strategies of New York focus on water supply, reducing water demand, water treatment systems, flood protection, and water quality. Measures of the Green Infrastructure Plan for the next 20 years include 1) water supply diversification, 2) strengthening water control, protection, and	Utilize groundwater and desalinated seawater, diversify the water supply network, protect water sources, control water consumption, improve water supply equipment, enhance land-use management and

7 Earth Hour started in Australia with millions of homes and businesses turning their lights off for one hour. Now it has grown to become one of the world's biggest climate change initiatives. People are all joining together in a global effort to show that it's possible to take action on global warming.

City	Climate change impacts	Adaptation measures	Specific practices
	than in other areas of the globe between 1900 and 2005. The sea level rose by 26 centimeters from 1920 to 2005. The number of high temperature days increased, demand for the water increased, river flow has decreased, and urban water supply is insufficient. The total rainfall from rainstorms is higher, which increases the probability of geological hazards, requiring higher reservoir capacity and dykes with higher flood capacity. Sea level rise raises flood risk and challenges urban sewage and drainage systems, especially on the coast.	conservation, 3) urban flood coping (including basements, sewers, subway), growing water retention systems and modifying the drainage design standard (rain water collection, natural retention, sewage. operation), flood diversion, 4) Water treatment systems: supply water with different water quality standards during the extreme weather, enhance water purification equipment, increase sewage capacity, develop better design standards, 5) Responding to the decline of water quality: adopt water-saving measures, upgrade water treatment systems and reduce runoff.	surface water purification, improve water quality control. Increase coastal dam construction standards, add submersible pumps and protective barriers, add emergency drainage equipment, control storm water diversion to a fixed area during storms.
UK (London)	London is a financial center and has an important trading port. The city faces	Geographical information Systems (GIS) are used for flood modeling, sharing the information and	London adopts proactive policies to slow climate change impacts,

City	Climate change impacts	Adaptation measures	Specific practices
	diverse climate change impacts such as higher temperatures, increased seasonal rainfall, drier summers, and more humid winter. It is predicted that the sea level will rise by 90 cm and the heat island effect will increase. Sea level rise will shrink salt marshes.	raising the public's awareness of floods. For drought, balanced supply and demand of water is sought by reducing water consumption, enhancing drought adaptation and establishing drought emergency plans. In order to control the impact of high temperatures, London increased green space and vegetation coverage areas, with a total target of 1000 acres. Climate risks are coordinated with public health organizations and infrastructure planning, mainstreaming climate risk into operational risk management and planning.	including low-carbon pathways to ensure energy supply, energy conservation, and pollution reduction in the private and public sector, such as low-carbon buildings, traffic, and industrial zones.
Japan (Tokyo)	Tokyo is an important coastal city in the Pacific and the core of the metropolitan and urban agglomeration of Japan. Tokyo's main climate threats are reduced food and drinking water	Tokyo connects with a long coast along marine reefs and imports many resources from others countries. It is very sensitive to sea level rise. Tokyo proposed a reduction of 25% of carbon emissions from 2000 levels by 2020 and to	Tokyo is developing a new type of urban economy, the low-carbon city, and a low-carbon technology concept, reducing environmental impacts and

City	Climate change impacts	Adaptation measures	Specific practices
	because of climate change and disappearing useable land due to sea level rise.	encourage a low-carbon and low-energy consuming society. The city makes use of new energy sources (solar, urban waste heat, etc.) and building standards such as urban green to improve the local micro-climate.	enhancing the charm of the city to attract more business investments. Tokyo strives to be the pioneer in low-carbon urban development in order to be more competitive in the global context.
Australia (Sydney)	Sydney has advanced infrastructure, economy, trade, finance, tourism and culture, and it is the largest city and harbor in Australia. Rising temperatures and changing hydrometeorology have a long term influence. Rising sea level causes coastal erosion and land degradation.	Sydney adopted measures for greenhouse gas emissions reduction and water conservation according to the corresponding provisions of the Kyoto Protocol. It strives to become the first carbon neutral city in Australia. Carbon emissions in 2050 are targeted to be 70% lower than 1990 levels. 25% of electrical energy is targeted to be produced by renewable resources in 2020, water consumption in 2015 is targeted to be the same level as 2006; urban waste recycling will reach 66% by 2014.	Sydney adopted a series of measures to reduce its dependency on combustion power, convert to low-carbon energy resources, and promote green energy-saving facilities. It promotes social strategies such as the National Cycling Strategy and "Earth Hour," with the aim to become an environmental protection pioneer.

City	Climate change impacts	Adaptation measures	Specific practices
Italy (Venice)	Venice is located 4 km from the shore in shallow water with an average depth of 1.5 m. It is an important harbor in the northeast of Italy and a famous water city. The development of the city is affected by rising temperatures, precipitation changes and sea level rise, which is main threat to Venice's development, economy, and security. Due to increased intense precipitation in the last few years, the city has suffered seasonal floods and large areas suffered sea water intrusion. In November 2008, Venice experienced the most powerful tide in the past 20 years, which initiated investment in a new flood control system, sea	Currently, the "Moses" project supporting land reclamation and flood control has had positive results.	

City	Climate change impacts	Adaptation measures	Specific practices
	tide monitoring, and an early warning system. The “Moses” project, a tide gate, was constructed between Venice and the sea. The project can control tidal energy and take advantage of inland sand deposits to cope with the challenges posed by rising sea levels.		
Singapore	Singapore is an island city in Southeast Asia and one of the region’s major economic hubs. Economic activities, especially related to the service industry and financial sector, are very sensitive to disaster risk and climate change.	Singapore’s goal is to develop a low-carbon economy, a clean environment, and clean air strategies. Its carbon intensity was reduced by 22% from 1990 to 2004 by increasing energy efficiency and using clean energy. The overall target is to reduce carbon intensity by 25% from 1990 to 2012. Singapore established a national energy policy framework to maintain a balance between economic competitiveness, energy security, and environmentally sustainable development.	Singapore’s Civil Defense Force established early warning systems, protection systems, and rescue systems; it leads natural disaster prevention and rescue work; it created strict building designs and construction standards.

City	Climate change impacts	Adaptation measures	Specific practices
South Africa (Cape Town)	Cape town is the second largest city in South Africa and faces severe droughts, storms and other extreme events. The water supply is critical, biodiversity is degrading, and the frequency of forest fires has increased. Coastal erosion is sever due to sea level rise. Infectious diseases such as dengue fever will increase due to warmer temperatures and increased flooding.	Cape Town has taken special measures to deal with the water supply, storms, and forest fires. In the coastal area, the government has developed a shoreline management plan and enhanced climate adaptation capacity to air and dust pollution, and reduced traffic. In the field of health care, the government has focused on strengthening infrastructure and construction of public health facility. It plans to further look at the insurance industry, banking industry, transportation industry, and communications infrastructure.	Cape Town has taken comprehensive countermeasures including a disaster management plan called "Disaster Management Behavior and National Disaster Management System." Risk assessments for public infrastructure are conducted. The government will pay more attention to the maintenance of ecological diversity and coastal area management.

Data sources: climate data web sites, city climate action reports and "Climate Change Adaptive City Entry Guide" and "Global Climate Change and Vulnerability Assessment of Estuary City—A case Study of Shanghai."

Many countries, including developing countries, have started researching potential impacts of climate change and have formulated "National Adaptation Action Plans." Some cities have taken the lead and formulated "Urban Adaptation Action Plans." These policies and experiences could be references in China. Cape Town, in the relatively developed area of South Africa, has drawn urban disaster and recovery plans to provide people with information on climate change.

Large cities in China in the Yangtze River Delta, Pearl River Delta, Bohai Sea, Shanghai, Guangzhou, and Tianjin have provided good examples for other cities. Shanghai has built coastal, island, and inland water observation sites, established a tidal warning and emergency system along the coast, strengthened coastal area's sustainable economic development management, bolstered public participation, and enacted environmental protections, including salt water intrusion control. Shanghai has also made evaluations and estimates of climate change impacts and of extreme weather events to provide scientific support for coping with climate change and sea level rise in Shanghai and the Yangtze River Delta.

New Orleans is the largest city in Louisiana and an important port and industry city in the United States. Because the elevation is below sea level, it faces the challenges of sea level rise and of sea water temperatures. Over the past century, New Orleans has lost a great share of its wetlands because of large-scale oil and natural gas exploitation and the construction of embankments and channels. In 2005, Hurricane Katrina forced the local government and the state government of Louisiana to change protection plans from "possible needs" to "necessary needs." They developed practical and effective strategies, implementing eleven natural and artificial defense lines in the Gulf of Mexico to prevent storm surges and protect wetlands. Its experience can provide useful examples to other cities in estuaries.[8]

B *Differences in Climate Change Adaptation in Developing and Industrialized Countries*

Coping with climate change in developed countries started earlier and important measures have already been implemented. For example, London and Sydney have already started evaluating new risks. These cities have also developed targeted adaptation measures, including specific industry actions. Singapore has found novel housing solutions by providing economical and practical commercial housing, enabling emboldened defenses against climate-related disasters. Venice has recently built new beaches, expanded the coastline, constructed new dykes, and repaired the ancient infrastructure. In developing countries, adaptation to climate change is very slow. Many cities face many severe problems such as underdeveloped infrastructure (roads, tap

8 Wang Xiangrong, Wang Yuan, *Global Climate Change and Vulnerability Assessment of Estuary City—A case Study of Shanghai*, (Beijing: Science Press, 2010), 36–47.

water supply, sewage and drainage systems, power supply), low-grade urban social service (such as sanitation and education), weak organizational capacity, including research, law enforcement, and generally low education levels. Urban low-income groups will suffer from climate change impacts in developing countries disproportionately.

Whether coastal cities adopt adaptation measures to cope with climate change will influence the lives of billions of people. Climate change and disaster risks call for urgent action. Urban planning and development should incorporate the construction of adaptive disaster prevention and disaster reduction capabilities. Future risks have to be considered as cities develop and expand. What most cities in developing countries need is not only adaptation planning but also development planning.

IV Challenges and Prospects for Climate Change Adaptation in Coastal Areas

A *The Main Challenges Faced by Developing Countries*

Coastal zones are important for society and for economic development. Global warming and accelerated sea level rise will greatly raise the basic risks in coastal areas. Many governments and international organizations are not yet focused on urban adaptation. Climate change adaptation in urban areas of developing countries faces the following challenges:

1 Limited Funds

The investment ability of cities of developing countries is comparatively limited. The funds to cope with climate change, especially current adaptation actions and disaster prevention and reduction projects, are limited. Funds to cope with future risks are more difficult to acquire and manage transparently in developing countries with weak governance structures.

2 Rapid Urbanization Aggravates Infrastructure Deficiencies

Developing countries, including China, are experiencing very rapid urbanization. Because of the large increase in urban population and property, the city expands constantly. The current disaster reduction capacity has kept pace with urban development, while danger zones and risk areas are growing, and infrastructure is already insufficient.

3 Disaster Risk Management is Limited in Coastal Cities and Disaster Prevention and Reduction Must be Enhanced

Cities lack public awareness and defense mechanisms against increasing extreme weather events caused by climate change. The risks of climate change and meteorological disasters have not yet been incorporated into the urban disaster prevention and reduction systems built into urban planning and construction.

B *Combined of Climate Change Adaptation and Urban Sustainable Development Strategies*

1 Coping with Climate Change should be Emphasized, Sifting the Comprehensive Strategy from "Climate Change Mitigation" to Include Adaptation

Governments and communities along the coasts should be aware of the influences and consequences of climate change, and should not focus all of their energy on energy conservation and emissions reduction technologies. They should also take active measures to adapt to climate change, protect wetlands, estuaries, and alluvial plains along the coasts, slow down coastal erosion and enhance natural defense capabilities.[9]

2 Climate Change Adaptation and Disaster Prevention and Reduction in City Development Plans and Emergency Systems

Climate change risks have to be considered in the construction of coastal cities by improving municipal engineering design standards, factoring sea level rise, and constructing environmental protection facilities, drainage systems, dykes, coastal roads, and ports. It is important to assess climate risks and vulnerability for different groups of society in coastal city development plans in order to develop monitoring and adaptation systems.[10]

3 Guide Low-Carbon Development and Raise Awareness around Disaster Prevention and Reduction

The public and the media need to actively address climate change. Their knowledge must be enhanced, and the public awareness of climate protection

---

9 Shi Yafeng, Zhu Jiwen, Xie Zhiren, et al. 2000, "The Environmental Impact Prediction and Prevention Measures of Sea-level Rising in the Yangtze River Delta and adjacent area [J]," *Science in China* (Series D), 30 (3): 225–232.

10 Li Cunxiu, Liao Guiqi, Tan Weibing, 1998, "The Climate Change Impacts on the coastal environmental and Countermeasures," *Journal of Guangxi Meteorology* 19 (3): 28–31.

should be raised. Scientific and technological support systems for climate change adaptation could actively boost the harmonious development of society and nature and the sustainable development of coastal cities.

Acknowledgment

This study was supported by the National Basic Research Program of China (973 Program) (No. 2012CB955903).

(This article was originally published in Chinese in 2012.)

Index